THE MONKEY WARS

THE
MONKEY
WARS

Deborah Blum

New York   Oxford
OXFORD UNIVERSITY PRESS
1994

Oxford University Press

Oxford   New York   Toronto
Delhi   Bombay   Calcutta   Madras   Karachi
Kuala Lumpur   Singapore   Hong Kong   Tokyo
Nairobi   Dar es Salaam   Cape Town
Melbourne   Auckland   Madrid

and associated companies in
Berlin   Ibadan

Published by Oxford University Press, Inc.,
200 Madison Avenue, New York, New York 10016

Oxford is a registered trademark of Oxford University Press

Library of Congress Cataloging-in-Publication Data
Blum, Deborah.
The monkey wars / Deborah Blum.
p. cm.   Includes bibliographical references (p.   ) and index.
ISBN 0-19-509412-3
1. Animal experimentation.   2. Primates as laboratory animals.
3. Monkeys as laboratory animals.   4. Animal rights.   I. Title.
HV4915.B58   1994
179'.4—dc20   94-12439

1 3 5 7 9 8 6 4 2
Printed in the United States of America
on acid-free paper

For
Peter
and
Marcus

ACKNOWLEDGMENTS

Shortly after I became a science writer in Sacramento, in 1984, I went on a routine tour of the nearby primate research center. I had a Mount Everest approach to it—it was there, I should see it. I didn't appreciate at the time how big the place was, thousands of monkeys camped in hundreds of cages at the California Regional Primate Research Center. What I remember from the tour is a small thing, one experiment, now defunct. Scientists were doing air pollution studies, testing the effects of toxic chemicals on the lungs of baby monkeys. They had built chambers—square glass boxes really—and they would put the monkeys inside and pipe the poisons in. The guide led me into the room and there was this box, empty except for two little monkeys. When we came into the room, they looked up. Both of them ran to the edge of the box. Their hands were pressed flat against the glass. They had tiny pink palms and long fingers. I stood there, looking at those little hands. And I thought, "I don't want to be here. I don't want to see this." I asked to move on.

Those monkeys, their wrinkled faces and small bodies, braced against the glass, came back to me when I was working on this book. I remembered that sensation of turning away, startling for a journalist trained to probe. It's one reason why the issue of animal research is such a haunting one. It forces us to see what we would rather not. Of course, we would rather have the drugs and surgical procedures and the wonderful medical advances without being told that so many hundred animals died for that information. And if we do know that animals are used, at least we would rather be spared seeing them bleeding in a medical test.

Yet, once you begin asking the questions, once you do know and see the animals, it's a hard issue to leave. At its foundation, the questions about animal research are fascinating, partly because they are questions about ourselves. At a time when humankind has so much influence over the world around us, we make choices in our care and use of other species. Those choices help define who we are, as masters, or stewards, or merely hapless residents, of this one small planet.

I had no idea, on that first visit to the California primate center, how many monkeys and apes I would eventually see. I've watched brain surgeries and colon examinations of monkeys. At the Laboratory for Experimental Medicine and Surgery in Primates (LEMSIP) in New York, I visited on a day when a little marmoset had chewed off the end of his tail, requiring a surgical amputation of the tip. The little monkey was absolutely furious. They gave him a local anesthetic and held him in a big leather glove during the surgery. He reminded me of an animated baseball in a mitt, if you could add on the tufted ears. The marmoset's voice, continually complaining, was just like Morse code: "Dit, dit, dit? Dit, dit DIT!" Easily translated into let me go! At Central Washington University, in Ellensburg, Washington, I stood baffled in a hallway, while a chimpanzee tried to talk to me in sign language. I finally grabbed a staffer to ask what the chimp, a male named Loulis, was saying. It was "Come, Play Chase." I felt ridiculously flattered. I visited the home of Shirley McGreal, head of the International Primate Protection League, in Summerville, South Carolina. She left me briefly in charge of a blind baby gibbon named Beanie, after warning that he could bite like a tiger. Beanie and I were sitting nervously on the sofa, but as soon as she left, he leapt away from me. Three times I had to haul him off the floor and back on the sofa, expecting any minute to have a finger taken off. That's what I remember most about him, not the dulled eyes or even the dark, graceful body, but the very sharp teeth.

They are fascinating animals in their own right, without tagging the science or the ethics to them. So are the people who work with them. People involved with monkeys—either in research or against it—are almost always opinionated and rarely shy about expressing those opinions. On many trips, I underestimated the number of tapes I would need to record an interview. I took to purchasing micro-cassettes whenever I saw them.

I was reminded, throughout, of how generous scientists can be, with their time and their ideas. I should emphasize that I did not visit every primate facility in the country, choosing instead ones that I thought would be representative, where I knew of interesting work or interesting people. Every place I went—the California center, the Yerkes Regional Primate Research Center and its field station in Georgia, the Tulane Regional Primate Research Center near New Orleans, the Wisconsin Regional Primate Research Center in Madison, and the Southwest Foundation for Biomedical Research in San Antonio and LEMSIP—people were open-handed both in showing me their work and their animals. The same was true at the individual laboratories I visited.

More than that, there were people at research institutions who called me, mailed me documents, checked to make sure that I hadn't missed an important discovery, allowed me open access to information in their own file cabinets. I owe particular debts of gratitude to Duane Rumbaugh, at Georgia State University; Jan Moor-Jankowski, at LEMSIP; Larry Jacobsen, the chief librarian at the Wisconsin Regional Primate Research Center; Cathy Yarborough, the head of public affairs at the Yerkes Regional Primate Research Center; Stacy Maloney, head of public relations at the Southwest Foundation for Biomedical Research; Jeff Roberts, assistant director of the California Regional Primate Research Center; and Peter Gerone, director of the Tulane Regional Primate Research Center.

For their patience in answering my many questions and repeated phone calls, I also want especially to thank Stuart Zola-Morgan, at the University of California, San Diego; Roy Henrickson, at the University of California, Berkeley; Roger Fouts and Deborah Fouts, at Central Washington University; Helen LeRoy, at the Harlow Primate Laboratory in Madison, Wisconsin; Robert Gormus, at Tulane; Jonathan Allan, at the Southwest Foundation for Biomedical Research; and Milton April, at the National Institutes of Health. I want also to thank Tom Gordon, at the Yerkes field station, for spending extra hours with me at a busy time, patiently and thoughtfully discussing primate research.

There were also people in the animal advocacy movement who went out of their way to provide information, records, and suggestions. In particular, Shirley McGreal, who allowed me to rummage through all her files; Christine Stevens, at the Animal Welfare Institute, who provided me with a wealth of information on animal welfare legislation and historical records; and both Suzanne Roy, at In Defense of Animals, and Steve Kaufman, at the Medical Research Modernization Committee, who patiently answered my many questions and backed their answers up with documents.

The idea for this book came from a series I did for my newspaper, *The Sacramento Bee*. My editors at the *Bee* supported my interest in monkeys from the beginning and were unfailingly generous in giving me time to work on both series and book. Marypaul Yates, Benjamin Weisgal, Maria Lapiana, Patrick Barreto, Daryl Metz, Janet Fullwood, Darcy Szeremi, Dawn Lewis, and Dana Kuehn listened to all my monkey stories with patience and affection and an always helpful detachment. My parents, Ann and Murray Blum, were less detached, but sent me every monkey story they saw for more than a year. The book owes a great deal as well to Claire Wachtel; to my very supportive agent, Suzanne Gluck; and to my editor at Oxford, Joan Bossert, who asked all the right questions. My husband, Peter Haugen, acted as critic, advisor, friend, and keeper of our 4-year-old son, Marcus, when I needed extra time on the computer, and deserves more credit than he will ever let me give him.

Sacramento, California   D.B.
March 1994

CONTENTS

THE
MONKEY
WARS

ONE

The Outsider

THE CASCADES RISE black and jagged against the silver sky of central Washington. As you drive eastward through the mountain range, coming from Seattle, they seem to push upward as if splintered from the earth. They belong to the Ring of Fire, the chain of volcanoes that wraps around the Pacific Ocean. Here, some of the volcanoes are still alive. They quiver faintly at their roots, promising more eruptions someday. Seismographs, placed by expectant geologists, mark the tiny rumbles and await the big ones.

For all that promise of volcanic fury, the mountains look deceptively solid. The peaks stand like castle battlements, sharp against the drifting clouds. The images they call up are all forbidding ones, of obstacles and barriers. Roger Fouts and his five chimpanzees live behind that rough wall.

Their home is the small town of Ellensburg, Washington, built on charm and century-old brick. Ellensburg is crammed with antiques shops and coffee bars. Of those two leading businesses, coffee comes first. Bookstores, clothes stores, gift stores featuring wooden cows and stenciled geese; all have a coffee bar. You don't even have to leave your car to get the daily caffeine fix. On Main Street, there's an old gas station, converted into a drive-through espresso stand. You pull up and stick your hand out for a cup steaming with hot latte.

The Chamber of Commerce area guide promises visitors the antiques, the coffee, the surrounding rivers and golden hills of the Kittitas Valley, even the bleak majesty of the Cascades. It also promises a chance to see psychology researcher Fouts and his "famous chimpanzees." That description of the scientist and his lab animals appears, in the guide, just above an advertisement for Ellensburg's Matterhorn Restaurant and Outdoor Beer Garden.

Roger Fouts has the relaxed manner of someone accustomed to attention, both friendly and hostile. He's a compact man in his early fifties, comfortable in blue jeans and sweaters. He has fair hair, just starting to gray, and bright blue eyes behind gold-rimmed glasses. His five chimpanzees are slowly sliding into middle age too. The oldest, a female named Washoe, is nearly 27. She is about halfway through the normal life span of a captive chimpanzee. There are two other females, Tatu and Moja, and two males, Dar and Loulis. Loulis, 15, is the baby of the group. They live where Fouts works, at Central Washington University, a leafy little school which offers bachelor's and master's degrees but no doctorate.

Fouts, Washoe, and Loulis arrived here in 1980, to begin a program in human-chimpanzee communication. The other chimps have been acquired since. All five animals are trained in human sign language. Fouts and his staff, which includes his wife, Deborah, spend hours watching the flicker of the apes' fingers as the animals talk to each other, to their toys, to themselves.

He hopes, by listening in, to penetrate into the essence of who they are, these big, dark creatures that can seem so like us. In the past 25 years, since he began working with Washoe, he has become more and more interested in quietly watching the chimpanzees talk to each other with sign language, using cameras set unobtrusively in corners of the apes' enclosures, operating them by remote control. The cameras have recorded the chimpanzees' talking to themselves—sometimes even swearing to themselves. He has taped them finger-chattering to stuffed animals, which Fouts sees as a clear sign of imagination. He's watched them squabble, call each other names, comfort each other. Such activities are hints, he thinks, of a complex intelligence in chimpanzees, long underestimated by humans.

Washoe is famed for being the first of her species to master sign language, to demonstrate that an ape could take hold of human communication and use it. Her early sentences sounded like the exclamations of a 2-year-old child—"Listen dog"—when hearing a bark. But who had imagined that an animal, even a chimpanzee, could gain a toddler's hold on human language? Understanding words, at some level, isn't that unusual. Dogs do it. Birds mimic human words. Pet owners think it's cute. But for an animal to actually speak a human language, and know what it was saying—that surprised some people, startled some skeptics. Many of those skeptics were scientists.

When Washoe first startled the research community, the people working with her became famous too, at least in scientific circles. Sometimes, here in the rural backyard of the Cascades, it's hard to believe that the Ivy League universities used to call Fouts. His work was heralded in the best science journals, funded by the big federal agencies. It's almost like another man's life. Washoe is still famous, but now it is more as a matter of history. The calls have stopped. The hotshot journals, Science and Nature, have lost interest. Fouts lost his federal support more than a decade ago. He keeps trying to regain it and he keeps getting turned down. On the worst days, he wonders how he will find the money to feed and house his animals.

He thinks of it sometimes as paying for a sin, at least a scientific sin. How did Roger Fouts turn from rising star to researcher without a grant? Why did he end up in this charming backwater, tucked in the shadow of the Cascades? He made a big mistake, for a researcher. He fell in love with the animals he studies.

Fouts calls his chimps by name, celebrates their birthdays with gifts and treats, puts up a tree for them at Christmas. When people come to work in his laboratory, he tells them their job is to serve the animals, as cook, as butler, as chauffeur. "I do the best I can, but what I do is exploitative," Fouts says. "I'm still using the animals. But I drill into my students that the chimpanzees are not ours. They are God's."

Perhaps even that wouldn't have been so bad, if he'd just loved them quietly. But he's been outspoken about it, and at a time when animal research and animal welfare have clashed bitterly. With the rise of the animal rights movement in the 1980s—hostile, sometimes violent and accusatory toward animal research—scientists have felt forced to choose sides. It's "us," the defenders of modern medicine, versus "them," the forces of ignorance and darkness. Or, if you're on the other side of the ideological barricade, it's "us," defenders of innocent creatures, versus "them," the heartless, sadistic butchers. Across that divide, people hate each other. A scientist who crosses over can get himself labeled a traitor. Roger Fouts made a decision, years ago, that his first loyalty was not to uphold his profession.

"If you look at the research community, somebody has got to stand up and say the king has no clothes," Fouts says. "Chimpanzees aren't numbers; they aren't hairy test tubes. They are folks who are going to have major problems if we don't replace human arrogance with compassion."

It's harder, perhaps, to step back from chimpanzees than to hold a distance from other animals, even other primates. In laboratories, as in the wild, they are relatively rare and thus, special. Only about 1,500 chimpanzees are kept in labs in the United States. A disappearing species, they are expensive to buy and too valuable to use casually. American researchers use some 40,000 monkeys every year. By contrast, they use almost 15 million rats and mice. In such quantities, animals usually don't get names, just numbers etched in tags around their necks, tattooed onto their skin. "We make them the same as soldiers in a war," Fouts says. "Without names, they become faceless, lose their identity. It's extreme exploitation, the same as in the labs of Nazi Germany."

Chimpanzees stand out for other reasons, too. In the debate over animal research, they've been a focusing point in spite of their rarity. The fundamental ethical question of that debate is this: "Is it right to use, and even to kill another animal to answer an interesting research problem?" With chimpanzees, the question comes home, or at least very close, because they are so uncomfortably like us. Genetically, they are human beings' next-of-kin. The genes within their cells differ by a wavering fraction, perhaps 1.5 percent from ours.

Some of that difference is obviously in the outer wrapping—the skin covered with dark fur, the long, strong arms, the feet with their opposable toes for grasping. The face with its heavy lower jaw. The brown eyes though, are steady and aware. They look back at you as if they see you, as if they are thinking about you. "You know there's someone home," is how Roger Fouts puts it.

One of the scientific tests of conscious intelligence is mirror-recognition. When you look in a mirror, is there a "Hey, that's me" response? Humans pick up on that early. An infant in its crib is fascinated by the sight of its own face in a mirror. Of animals tested, it is chimpanzees that most clearly show that same jolt of recognition. To test this, researchers put chimpanzees under anesthesia. While the animals are unconscious, researchers color a brow ridge or an ear bright red. After each is awake, the chimps are shown their reflection in a mirror. In every case, the chimpanzee shows shock. Staring in the mirror, the chimps begin feeling around their heads, probing for the disfigurement. There is an obvious reaction of "That's me and I don't look right."

When similar tests have been tried on monkeys, the reaction is different. Monkeys seem to recognize the appearance of a monkey— some have actually prowled behind the mirror, looking for the rest of the animal—but they show no signs of dismay at their own bright-splotched faces. There's no clear sense that they recognize the face in the mirror as their own—unlike chimpanzees, unlike people.

Fouts thinks of his chimpanzees as people. He laughs at the way Washoe expresses her indignation, signing "dirty, dirty"—her favorite swear word—when she doesn't get what she wants. Or the way the two young chimps, Dar and Loulis, try to blame each other for fights, signing "Me good" when caught squabbling, pointing accusingly at their companion. It's clear, too, that the chimps think of Roger as the local soft touch. When caretakers pass by, the animals frequently sign for treats: nuts, ice cream, chewing gum, coffee. The request is often "Roger, ice cream," "Roger, coffee."

Yet Fouts, dispenser of treats, is all business about their care. He has fired caretakers who are casual about cleaning the cages. For all their charm, too, chimpanzees are big animals and strong ones; approached too quickly, too roughly, they can be dangerous. Fouts keeps strict rules about dealing with the chimps. "I may come across like a nice person," he says, "but cross my rules and you'll find out that I'm not nice at all."

In a way, the chimpanzee laboratory in Ellensburg is his own little kingdom. It is his show and they are his chimpanzees. He sees them every day. They coax him to give treats. They sign for him to "Come hug." The Cascades are not really a very good wall; they don't shut out the rest of the world. But, behind them, your own world, your own concerns and rules can become the Truth, with a capital T. Living there, knowing his animals so well, he has begun to feel more and more obligated to speak out on their behalf. He worries sometimes about developing a messiah complex, believing himself more important to the chimpanzees than he actually is. But when he hears of a chimpanzee sick and alone in another laboratory, it's always his own animals in his mind. It's like being the parent of a small child; you read about child abuse in the newspaper and it brings tears to your eyes. You think of your own small son in pain, bewildered and frightened. Roger Fouts sees Washoe, defenseless.

So, he minds when other scientists cut open chimpanzees on an operating table or inject them with an infectious disease. He has minded enough, in recent years, to take on his own profession, accusing his colleagues of indifference to the needless suffering of animals. He's gone past that point, working with animal welfare groups, criticizing work by other scientists. And finally, in 1992, he joined with animal advocacy groups in suing the U.S government, challenging the way that it regulates the care of captive animals.

If there was an invisible line in the sand—step over this and you are no longer one of "us"—Fouts was across it with the lawsuit. He did it knowingly and in anger. His affection for his animals has led him to alienation from his profession. It's as if there's no room for him, for a researcher who becomes too fond of the animals he studies. He cannot be respected because he cannot do brilliant research because he will not sacrifice his animals to a really interesting question. He feels the dismissal, that he's a nice guy, maybe, but too sentimental. It stings. "The arrogance in science is unbelievable. We (scientists) assume that what we do is beyond question. The Medieval church is a parallel, with the National Institutes of Health, rather than Rome, as the center. I seldom embrace the scientific priesthood myself, but I can see why it's appealing, to never question what is done, to think this is IT. But look at what science has become. Question it and do you know what the response is? It's 'Who do you think you are?' And that will change, maybe, when people die and we go onto a new generation of scientists."

IT CAME OUT OF nowhere, this gift for relating to chimpanzees. He'd grown up on a farm, in California's central valley, just south of Sacramento. He thought of animals practically. "They were our friends, but sometimes we had to eat our friends for dinner." As a budding psychologist, in the hot, dry summer of 1967, he wanted to work with children. He was also in desperate need of a job. He had finished his master's degree at Long Beach State University; what he admits was kind of a double major in psychology and surfing. He'd managed, in between wave-catching, to maintain decent grades. Decent, but they were not grades to lift a young scientist from Long Beach to Cambridge. Like Central Washington, Long Beach State didn't have a Ph.D. program. Fouts applied to nine programs in clinical psychology. They all turned him down.

The University of Nevada, though, offered him a slot in its experimental psychology program and the possibility of a research assistantship. Fouts drove from his California home, east through the soaring rock and deep pines of the Sierra Nevada mountain range, down into the desert town of Reno. He brought his own cheering section—his parents, his wife, his 6-month-old son, Joshua. Yet, despite the fan club, despite his need of the work, he was dubious before he ever stepped out of the car. The Nevada assistantship just seemed, well, weird.

Two experimental psychologists, the husband-wife team of R. Allen and Beatrix Gardner, had a grant to home-rear a baby chimpanzee. The Gardners wanted to test an interesting theory. People had been trying, for years and without success, to teach apes to talk. The failure of those experiments had been dismissed as primate stupidity. The Gardners suspected something else. Perhaps the problem was chimpanzee vocal cords, that they couldn't form the proper sounds. If they could comprehend language, though, they might be able to pick up sign language. That was the plan, to teach their chimpanzee, named Washoe for Washoe County, Nevada, to talk with her fingers. The Gardners had a half-time research assistantship available for someone to help train the animal.

On a Sunday morning in August, Allen Gardner met Fouts for the interview. As Fouts recalls it, it was a case of mutual dismay. Fouts babbled nervously about his love of children and his desire to do clinical psychology. Gardner replied coldly that he preferred reasoned experiments. Fouts burst into praise of the influence of philosophy on science. As Fouts recalls, Gardner, glaring, said he didn't believe that science needed philosophy and that "if you learn that, it'll show you weren't worth anything anyway." Fouts began to wonder if maybe he wasn't meant to be a psychologist. Finally Gardner, grudgingly, offered to show him the chimpanzee. It was obviously meant as a sort of a loser's consolation prize: you're not going to get the job, but I'll let you look at our amazing animal.

The Gardners were trying to raise Washoe in a friendly environment, family style. They'd reasoned that a nurturing family was one of the key factors in developing language, that it encouraged people to communicate. So, like a good family, they took the chimp on weekend outings. Perhaps typical families didn't spend Sunday morning at the university's home economics department. But the white, one-story building was home to the university's nursery school. The playground was crammed with jungle gyms, swing sets, and sand boxes. Empty of children, it made a great space for Washoe.

Through the afternoon glare, Fouts could make out two adults and what he thought was a child, playing in the shade of a tree. Suddenly, the child began to hoot. It started racing, all four legs on the ground, scrambling toward the two men. Fouts did a brief double-take. He squinted harder, recognizing the chimpanzee just as she reached the four-foot-high, chain-link fence. As she scrambled toward them, Fouts marveled at the strength of the bond between Gardner and his animal. Then Washoe pushed herself off the warm pavement and went airborne, straight toward Fouts.

He put out his arms and thirty pounds of chimpanzee thudded against his chest. Washoe's long arms wrapped around his neck, her legs around his waist. They clung together for a startled moment, Fouts's surprised eyes meeting Gardner's over Washoe's dark head. "We weren't supposed to talk around her," Fouts said, "So I just smiled at her," and he could feel the smile tugging, wanting to be a grin, wanting to stretch into a laugh. Instead, he carefully handed the chimp over to Gardner. As he smiled at Gardner, suddenly, he knew he was in. "Washoe wanted me," Fouts said. "Gardner was caught."

Fouts still wondered what he was doing. He was still unnerved by the idea of working with the chimpanzee. His recollection of his first days with Washoe is of bare survival, for both of them. In one instance, a month later, he was alone with Washoe in her trailer. They'd had breakfast together and Fouts was washing the dishes while the chimpanzee watched. The cleaning supplies for the trailer were kept in a locked cabinet under the kitchen sink. Fouts opened it to get out the soap for the dishes. As he was putting the bottle back into the cupboard, Washoe grabbed the sodden dishrag, popped it her mouth, and hurried over to the bedroom for a game of "chase", one of her favorites. "I finally retrieved the rag, or should I say that Washoe tired of the game and gave it back to me," Fouts says. The chimpanzee went back into her bedroom to play with dolls. The fledgling scientist cleaned the dining room table and sat down to make some notes in the logbook.

He had forgotten the unlocked cupboard. Washoe, however, had not. She swung out of the bedroom, skidded to a stop in front of the sink, grabbed a bottle, and fled back into her room. Fouts tells it like this:

"I was on my feet and running. When I reached the bedroom I saw Washoe sitting on her bed chug-a-lugging a bottle of Mr. Clean. I screamed at her in terror. I thought that Mr. Clean was poison and there she was drinking it. My scream startled her. I picked her up and rushed her into the kitchen and sat her firmly on the table and tried to gather my wits. She sat there frozen with fright because of my strange, panic-stricken behavior."

Desperately Fouts tried to recall anti-poison remedies. The only one he could remember was inducing the victim to vomit. Right. He caught her head in his arm, opened her mouth, and stuck his finger down her throat. Where was her gag reflex? He couldn't find it. What was he doing with his finger stuck in a chimpanzee's mouth anyway? Then he remembered that antidotes were usually printed on the labels of toxic materials. He grabbed up the Mr. Clean bottle. Unfortunately, he was so panicked by that point, he couldn't read it. His eyes kept flicking back to the chimpanzee, checking for signs of imminent death. "All I could think was 'You've ended the project ... you let Washoe drink the poison and if she dies it's YOUR fault.'

"Finally I was able to focus on the label and read it. There was nothing there about poison or antidotes. I began to panic again, all I could think of was finding an antidote." Then he remembered milk, wasn't milk supposed to absorb some poisons? He rushed to the refrigerator and grabbed up some of Washoe's baby formula. By this time, though, the little chimp was so unnerved by Fouts's near-hysteria, she refused to drink. She jumped off the table and ran into her bedroom, slamming the door. Fouts tried once again to study the label. Slowly it began to dawn on him that there was no skull and crossbones, that he was holding a bottle of strong soap. He remembered his mother washing his mouth out with soap, in response to a fourth-grade profanity. Surely, she wouldn't have tried to poison him?

By day's end it became clear that Washoe was going to survive with only a bad case of diarrhea and that Roger Fouts was going to spend the day cleaning the trailer floor. "From my point of view, it was a small price to pay for my carelessness." He had not killed the project and the project, everyone was realizing, was going to be a very big deal.

People had known for a long time, of course, that chimpanzees could understand human words. Lots of animals can learn commands. Scientists kept thinking, though, that chimpanzees might be able to do more, to talk back. In the 1930s, a young couple, Winthrop and Louise Kellogg, decided to test the idea by raising their 9½-month-old son, Donald, with a young chimpanzee. They wondered whether an ape, raised like a human, could learn to think like one. As Donald's companion, they chose a 7-month-old female chimpanzee, named Gua. Then they simply raised the pair like children. Tracking Gua's development is enough to make any parent of toddler-age children grateful for the relative slowness of human physical development. Gua could open a swinging door at 8 months, unhook window screens at 10 months, and by 13 months could unlock the front door. Donald was still learning to stagger-walk across a room.

Gua also picked up on language faster than Donald. Some two months into the study, she could respond to seven different requests; Donald, though older, acknowledged only two. The Kelloggs continued the project for nine months; they ended it only when it became clear that Donald was going to leave Gua behind. By that time, he could respond to 68 commands and she to 58. What seemed most impressive was Gua's quick grasp of the world around her. She turned on lights and water faucets and turned keys in locks. Winthrop Kellogg wasn't dazzled by Gua's language abilities but he became a great admirer of her physical skills. Gua, he wrote at one point, had the mind of a 1-year-old, the agility of a 4-year-old, and strength in some ways surpassing that of an 8-year-old.

The Kelloggs' experiment attracted more interest. Another pair of psychologists, Keith and Cathy Hayes, adopted a 3-day-old chimpanzee, Viki, and raised the ape as an only child. Viki was with the Hayeses until she died of viral encephalitis at 7. Like Gua, she was remarkably quick at figuring out the mechanics of things. She rapidly learned to use a needle and thread. On one memorable day, she unlocked the Hayes' car, climbed in, turned on the ignition, and proceeded to drive off, with Cathy Hayes chasing hysterically on foot. (Luckily, the car was barely moving. Cathy was able to catch it, jump in, and hit the brake.)

The Hayeses were sure Viki was smart enough to talk. They labored to teach her language. At 10 months, she was able to puff out a breathy "aahh" sound; the Hayeses molded her lips until she could say, "maah, maah"—Mama for Cathy Hayes. Eventually, she could also puff out "papa," "cup" and "up." A four-word vocabulary, however, had its limits.

You might make the assumption that chimps were linguistic idiots. Or you might, as researchers finally did, realize the genetic differences between humans and chimpanzees accounted for another critical difference. Chimpanzees do not have the kind of vocal cords that can make words. It wasn't that they didn't understand the language, they were physically unable to make the proper sounds. So the Gardners, in Utah, decided to embark on Project Washoe.

They took advantage of the fact that Washoe seemed to be naturally theatrical. She gestured constantly with her hands. Without being taught, she would extend her hand, palm up, when begging for something. The Gardners, using a suggestion of Fouts', expanded on those natural movements. They shaped her hands with their own while showing what the gesture meant. A crooked finger, for instance, would bring a person to her. It meant come here. Washoe learned 34 words the first year; 85 by the third; 132 before the project ended. The list included "come," "give me," "more," "up," "sweet," "open," "tickle," "go out," "hurry," "here," "listen," "toothbrush," "drink," "hurt," "sorry," "funny," "please," "food," "eat," "flower," "cover," "blank," "dog," "you," "napkin," "bib," "brush," "hat," "I," "me," "shoes," "smell," "plants," "clothes," "cat," "key," "baby" and "clean."

What was even more interesting about Washoe's early development was that she began, spontaneously, to string the words together into simple sentences. "Give me tickle" was one of her first, and it came before the experimenters had ever asked her to tickle them. (Chimpanzees, apparently, really enjoy being tickled and Fouts tells the story of another signing chimpanzee, later in his career, who repeatedly asked a visitor to tickle him without success. Finally, the big chimp poked the man in the chest and signed "Tickle me, you nut.") Washoe also came up with "Open food drink" to indicate the refrigerator; "Open key" for a locked door, "Listen dog" when hearing an unseen dog barking; and "Listen eat" at the sound of an alarm clock signaling mealtime.

The Gardners insisted that the signs be clearly made or they didn't count them: "dog," for instance, was repeated slapping of the thigh; "listen" was touching an index finger to the ear. Even so there were plenty of critics, people who wondered if the chimps really knew what they were doing. The sign for "listen" was merely touching an ear; what if the chimpanzee was actually scratching an itch? Critics suggested, too, that the scientists were somehow cuing the animals, making helpful gestures with their own hands. The challenge of how to make chimp language believable was taken up by another scientist, Duane Rumbaugh, an Iowa-born psychologist working in Georgia.

Rumbaugh was convinced from the beginning that the subject was sensitive—chimpanzees edging into the distinctly human realm of language. He worried, too, that the research was soft, needed a harder edge to be convincing. He thought the field would survive only on solid, tough-minded analysis. He didn't want a cuddly, keep-the-chimpanzee-at-home approach. He wanted a clearly measurable approach, cold data clicking directly into the memory of a computer.

In 1971, Rumbaugh was a staff scientist at the Yerkes Regional Primate Research Center in Atlanta. Along with an engineer named Harold Warner, he developed a keyboard on which each of a large number of keys was embossed with distinctive geometric symbols. Each symbol was called a lexigram. The idea was to train a chimpanzee to type out sentences on the keyboard, asking for food, drinks, hugs. A proper sentence would bring response, either from a person or a vending machine, linked to the keyboard.

The project, done with scientists from the University of Georgia and Georgia State University, was called the Language Analogue Project. The chimpanzee was named Lana, combining the "L" from Language and the "Ana" from Analogue. As with Washoe, Lana startled her trainers. She quickly learned to arrange symbols into "chimpanzee sentences," asking her trainers to come and tickle her, tapping the keyboard to ask for food. "Please, machine," she would begin to convey her need for a piece of banana, a glass of soda. Lana would studiously tap out the symbols: "Please, machine, make window open." The right symbols got her what she wanted. Ask for a banana and an automated door would slide open, revealing slices of fruit. If she didn't know the right word, she'd describe it: "Banana which is green" for cucumber. She also learned to demand objects outside the vending machines, admiring a "box which is purple," actually a purse carried by a visitor.

She could talk with trainers—and argue with them too. In one much-quoted exchange, one of Lana's trainers had taken the regular monkey chow out of the vending machine, replacing it with chopped cabbage. Five times in two minutes, she asked for the chow. Each time, the trainer assured her it was in the machine. Finally, the chimpanzee called his bluff.

LANA: Chow in machine?

TRAINER *(lying):* Yes.

LANA: No chow in machine.

TRAINER: What in machine?

LANA: Cabbage in machine.

TRAINER *(confessing):* Yes, cabbage in machine.

LANA *(exasperated):* You move cabbage out of machine.

By the early 1970s, Lana, Washoe, and their laboratories had brought chimpanzee language studies into respectability. More than that, it was attention-grabbing stuff. Talking chimpanzees were irresistible to the public. Federal agencies were intrigued by the new hints of primate intelligence. The scientists who did the work stood in a virtual shower of grant money.

The University of Oklahoma in Norman, which already had an Institute for Primate Research, decided that language studies would give it a new focus. The institute was run then by a longtime psychologist named William Lemmon. Lemmon had begun by studying maternal and sexual behavior in primates. He was willing, though, to pursue another promising field. He recruited Roger Fouts, who by then had completed his doctorate. Fouts and his family arrived in Oklahoma in 1971, along with Washoe. The Gardners continued their studies with other chimpanzees.

Oklahoma was a culture shock, for Fouts and for his chimpanzee. Washoe was 5 years old. She had never seen another chimpanzee. She ranked the animal world on a scale beginning with humans and ending with bugs, despised little creatures which she would flick away with her fingers. She had no idea there were others like her in the world.

"Imagine being raised in species isolation and never meeting your own species until you were 5," Fouts says. They flew into Norman with Washoe heavily sedated for the trip. Lemmon insisted she go directly into a cage, in a room filled with other caged chimpanzees. It wasn't a matter of seeing her first chimpanzees; it was being tossed into the crowd. There were then 25 chimps at the institute.

That was Fouts's first real experience in going up against the system, against the established way of doing animal research. He didn't like it. He was a brand-new employee and hesitant to push too hard. Over his protests, Lemmon took away Washoe's blanket, a clingy security symbol, as the chimpanzee slept. When she awoke, she was in a cage with concrete floor and metal bars. Fouts's one victory was being allowed to stay near her. He was pacing outside the cage. As Washoe struggled to her feet, the others started screaming at her, banging on their cages. It's hard to say who was most appalled at that moment, Fouts or Washoe. "What are they?" Fouts asked Washoe, gesturing to the screaming animals around her. "Bugs," replied Washoe. "Black bugs."

She soon learned otherwise. They both did. The up side of the move was that, gradually, Washoe began to accept herself as a chimpanzee and to accept the other chimps as well. Fouts, too, discovered that he had a gift for getting along with chimpanzees beyond Washoe. It wasn't just that he liked them. He felt that he saw them not as a caged mass of apes, but as individuals. He took pride in his ability to meet the animals on their own terms.

He was willing to go a lot further than most. One of Lemmon's chimpanzees was a nasty-tempered old male named Satan, a former zoo chimp who loathed humans. "He'd really been through the wringer," Fouts says. "He'd picked all the hair off his arms; you could see his muscles. He'd picked the hair off his face, too, and it really gave him a satanic look. He threw feces at anyone who came near him, which no one liked too much. One day I went in there and he started to display, you know, posture in a threatening way. He began scooping up feces and throwing them at me. I thought, don't flinch, don't flinch, he's entitled to be unhappy with people." The slimy chunks of feces hit Fouts in the face. "And I just stood there. He threw until he emptied out his cage and then he was desperate. So he started to spit water at me."

Chimpanzees are notorious spitters; in laboratories, they've been known to watch a scientist walk by, to fill their mouths with water in preparation for his return trip and then to spray him the minute he reappears. Loulis, the youngest chimp in Fouts's group, can spit a stream of water eight feet. He takes apparent pleasure in soaking unsuspecting newcomers to the lab. Satan drenched Fouts so thoroughly that he was polished clean by the time the spray stopped. "Then, he started to talk to me," Fouts says. The scientist is sitting in his office at Ellensburg, telling this story, surrounded by the trappings of a university professor, the computer, the stacked books and papers. As he describes Satan's actions, he bursts into an imitation of a conversing chimpanzee, a series of soft hoots, suddenly and startlingly nonhuman in sound. "So he said "hoo, hoo" to me and I hooted back," Fouts said. "After that, we were the best of friends."

It became clear very early that Lemmon despised this sissy attitude toward animals. Although Fouts remained at Oklahoma for almost ten years, it was in an atmosphere of increasing hostilities. Lemmon died in the late 1980s, but Fouts has yet to forgive him. "The director was from the old, but still-popular school of captive treatment that explicitly held that humans had to dominate the animals they owned and the best way to do this was to arbitrarily mistreat them."

Fouts has many stories, all ugly, about the way Lemmon treated animals. Lemmon bought Doberman pinchers to patrol the institute but he used to enjoy intimidating the chimpanzees with the snarling dogs, Fouts recalls. That stopped the day Washoe backhanded a Doberman, slapping the dog head-over-heels backward. Not that Dobermans had such an easy time either. One was killed while exercised by being tied to the bumper of a moving car.

Fouts and Lemmon were clearly incompatible. They stayed together a decade mainly out of mutual ambition. Lemmon wanted the Oklahoma program to shine; Fouts was its star. Psychology students came to Norman because of the national reputation of Fouts and Washoe—and a shared fascination with the study of chimpanzee communication. Fouts was young enough, prominent enough, and still accepted enough by the research establishment to be recruited by the important schools. In 1973, Yale University flew him East for an interview. By that time, though, Roger Fouts had strong opinions about what he considered acceptable treatment of his chimpanzees. They were unconventional opinions, as divergent from Lemmon's they as could be. Yale didn't meet his standards.

The monkey facilities were in a basement, neat, clean, white tiled rooms. "There was no place for the animals to climb," Fouts recalls of his trip to the university in New Haven, Connecticut. "I asked, what about blankets? And they said, oh, no, they would clog the drains. I said what about toys? They said no, they would get feces on them. I was standing there listening to lab techs talk about moving a monkey, and it had a number, not a name. And I said, well, give me a couple weeks to think it over. And they said, 'You're not going to take it, are you?' I asked why they said that. They said, 'No one asks Yale to wait a couple weeks.' "

In 1980, Fouts gave a lecture at Central Washington University. It was a continent away from Yale, a world away in scientific stature, but that school was also in search of a star. University officials asked what it would take to bring him out. Fouts's first response was a joking, "You can't afford me." They argued that they could, pointing out that they already had animal research facilities. The animal lab (which Fouts finally moved out of in 1993) was small, but he thought it an improvement over the Oklahoma facility. "If you'd visited in Norman," he says. "You'd have had the impression I was studying cockroaches. I went back to Oklahoma and decided it would be a big step up for the chimps in terms of their cages. It was a step down for me because there's no Ph.D. program here, only a master's. But on the other hand, I'd be working with undergrads and they're not so arrogant. You have to be humble to work with the chimps and meet them on their terms."

He and his wife talked it over and decided to move. "The dean at Oklahoma was upset. He said stay here a few more years and you'll be offered a job at Harvard. I said, I don't necessarily want that. I've already turned down Yale. If I'd wanted big biomedical I would have made a different choice."

And so, he went to Ellensburg, taking Washoe and Loulis with him. The Gardners, who had continued their own chimpanzee studies, later shipped him Dar, Tatu, and Moja. He was then—as he is now—Central Washington University's most visible, attention-getting scientist. Almost as soon as he arrived, though, everything else started to change.

When he moved to Washington, Fouts brought with him a fat federal grant, underwriting his continuing studies of sign language in chimpanzees. By the end of the year, the grant had ended. The National Institutes of Health did not renew it. Fouts recalls that the grant officer encouraged him to apply, instead, for a grant to conduct medical experiments on the animals. "I said, 'I study animal behavior, I'm not going to take money to harm the animals.' And that was the end of that." He has tried, repeatedly, to get new government support for his own style of research. He applies every year to NIH and to the National Science Foundation. Every year, they say no.

At first, the money issue wasn't personal—as it seems now. By the early 1980s, primate language/intelligence programs were no longer trendy. There was a growing attitude, at that time, that chimpanzees were simply a high-class version of trained dogs; they could learn the commands but they had no real grasp of language. If they signed to themselves, they must be mere copycats. "People just didn't want to believe it," Fouts says. "We've still got that arrogant, Mr. Rogers notion—you're special, we're special. Listen to

Mr. Rogers. No one wants to be ordinary. We want to be different from chimpanzees. We want language to be ours."

He and others took up the challenge. Many credit Sue Savage-Rumbaugh, a primate intelligence specialist in Georgia, for best proving the critics wrong. Savage-Rumbaugh is Duane Rumbaugh's wife and research partner. Her husband credits her with keeping chimpanzee language work going, and keeping the money flowing in their laboratory at a critical time.

In 1980, the year of the backlash against studying communication skills in chimps, Savage-Rumbaugh was one year into a five-year grant to study ape-language. That money, says Rumbaugh, carried them through the dry spell. And before it was gone, Savage-Rumbaugh had shown conclusively that chimpanzees do, indeed, understand language as a complex communications system. Rumbaugh describes her breakthrough study with envy. "I'd give four pages of my résumé to have come up with it myself," he says.

Savage-Rumbaugh wanted to take on the question of whether the chimps understood the classification of words—a foundation of human grammar. She decided to ask them directly. She was working with two male chimpanzees, Sherman and Austin, trained on the computer keyboard. She showed them 17 symbols—or lexigrams—each representing either a food or a tool. Then she asked the two chimpanzees to classify the symbols, group them according to whether they were a food or a tool. The chimps made one error only. Sherman called a sponge a food. "But when one acknowledges that Sherman literally ate a lot of sponges, which were used to soak up soft drinks ... one can understand why for him (but not for Austin), a sponge was a food. It was something he ate—after using it as a tool," Rumbaugh notes. Not so different from the way a person might use a potato chip as an edible tool for scooping up onion dip.

"It was a bench mark study," Rumbaugh said. "It showed that the animals could comprehend word meaning. It was incredibly straightforward. They just got it." Savage-Rumbaugh went on to work with the other, even rarer species of chimpanzee. (There are only two, the common and the smaller bonobo chimp.) A bonobo named Kanzi proved even quicker at understanding the way human language works. The two Rumbaughs have collaborated on many of the studies. In recent years, Rumbaugh has also expanded his own work into primate intelligence, questioning the animals' ability to understand numbers, do simple math, calculate.

The Rumbaugh laboratory has funding from several government

agencies, including the National Institutes of Health and the National Aeronautics and Space Administration. What Rumbaugh remembers, though, is how hard critics tried to brush off the word classification study when it first appeared. "You would have thought that linguists, people interested in the origin of language, would have embraced it. And it was just ignored. Nothing is more frustrating or disappointing in your profession than to have a fine piece of work ignored because it doesn't fit in with the belief of the day. Scientists are human in their reactions—sometimes unfortunately."

Federal officials have two basic explanations for why scientists like the Rumbaughs are still funded and Fouts is not. One is that he's at a backwater little school that doesn't support a Ph.D. program. "If he wants money, why doesn't he move to the University of Washington at Seattle," complained one administrator at the National Institutes of Health.

The other is that his research isn't that good, his proposals don't measure up against the elegant work done by researchers like the Rumbaughs. He's too close to the animals; too careful with their feelings. He won't use keyboards because they're unnatural to a wild animal. He doesn't push them to acquire human capabilities, analyzing numbers, staring at computer screens. He tends to step back and observe, rather than pushing them to perform. Reports from his laboratory include watching chimpanzees sign to themselves, speculating about whether that means they talk to themselves, as we do. "Observation is legitimate science," Fouts says. "It's letting the animal tell you what's important. Maybe it's too humble for some people."

The perception among researchers is that even a brilliant scientist won't get money if he's made NIH angry. So, the chances of a Roger Fouts doing neat, straightforward work are diminished from the start. Further, Fouts has obviously not been sitting quietly in Ellensburg, polishing up his chimpanzee experiments, trying to make the federal government happy. His friends, increasingly in the animal advocacy movement, believe that NIH has cut him permanently out of the loop because he speaks out. It's more than a suspicion. Christine Stevens is convinced of it.

Stevens—who heads the Animal Welfare Institute, a private, nonprofit group in Washington, D.C.—is a pragmatic worker for better care of lab animals. She knows federal health officials well because she believes in working within the system. She's widely credited as one of the most influential forces in the country's national animal welfare laws. She has counted Fouts as an ally for a long time and she has watched his career slide from hotshot Yale recruit to impoverished scientist. "Roger talks back to the people at NIH. He doesn't let things slide," she says. "He's a wonderful scientist. But will he get federal funding? I doubt it."

Deborah Fouts also thinks NIH's disapproval has hurt the lab she manages for her husband. "Roger's very vocal," she says, "and I didn't realize the extent of the hostility from other scientists that would result from that. He waves the flag a lot and that's made it more difficult to get grants."

That perception of the all-powerful, never-to-be-crossed NIH may be, as the agency insists, all perception. The NIH response is that all grants are awarded strictly on merit, after being judged by a panel of expert scientists: "People get burned at us and we're an easy target," says Milton April, who heads the chimpanzee program for NIH's Center for Research Resources. "The fact is, with someone like Roger Fouts, that they are judged by their peers. People think of us as a kind of master controller, pushing buttons everywhere. That's not true."

The closest Roger Fouts has come to documenting that his pro-animal stance works against him is not through NIH, but another agency. In the world of federal science grants, a rejected researcher is allowed to ask for the criticisms. The records are kept in triplicate—white, pink, and yellow. The researcher gets the pink copy. The critiques are known throughout the scientific community as the dreaded "pink sheets." Roger Fouts has one, from the National Science Foundation, which at least reinforces the idea that the system is not blind to animal research politics: After evaluating the grant proposal, the reviewer wrote "The investigator [Fouts] is a member of a number of organizations which oppose animal research of all types. This seems incongruous given the present request for funds." In addition to working with Stevens's group, Fouts belongs to Psychologists for the Ethical Treatment of Animals, a group of professionals frequently mistaken by researchers for the far more radical organization, People for the Ethical Treatment of Animals.

Since the federal grants disappeared, Fouts has sometimes felt an unwelcome tenant at his university. Central Washington pays him a solid salary, in the mid $50,000 range, but contributes only a $1,000 a year toward operation of his laboratory, he says. He's been told that other scientists resent the public attention he gets and that administrators regard him as notorious rather than famous. He's looked outside for support, creating a nonprofit foundation, Friends of Washoe. It has some strong allies. Jane Goodall, the famed chimp researcher, lobbied the Washington legislature in the lab's behalf, helping the Foutses get a $1 million toward a new chimp building. Fouts's office is plastered with photos of Goodall with him, his students, his chimps. Friends of Washoe has grown from 60 founding members to more than 400. The foundation generates about $100,000 a year, about half the operating funds that Fouts thinks necessary. The Washoe foundation receives contributions from other foundations, from individual members, and from their small shop. The Foutses sell T-shirts that say "Save Our Sibling Species." They sell stationery and prints of blobby chimpanzee art work to supplement their funds. Some days it's not nearly enough. Some days, it's something to fume about.

Deborah Fouts, who manages the lab, planned to be a psychologist too. She earned a master's degree at Central Washington University. The doctorate, though, has been out of reach. She would have had to leave Ellensburg to get it. She could have done it, except for the chimpanzees. There's not enough money for caretakers in her absence. "I love Roger," she says. "I've been his wife for 28 years. But I haven't had the profession I thought I would."

In 1993, she received a salary for the first time in 12 years, under a special grant from the state legislature to help open the new building. She was shocked at how differently people treated her when they went to professional meetings, the difference between being Roger's pretty, dark-haired wife, who volunteered in the lab, and a paid professional. People treated her like "someone" all at once, treated the lab as a real operation instead of something slightly amateur, run mostly by volunteers.

"Roger has always said the chimps come first and we've raised them like family members. When our daughter Hilary (the youngest) was small, she'd share her ribbons and treats with the chimpanzees. I think we've helped chimps in general but I'm not sure, in the long run, that we've really helped ours. Things are still hand-to-mouth. Of course, my first concern is these guys. But I temper things more. I would tend not to be as vehement as Roger. He's very passionate about the chimps. These guys have an uncertain future, though, because of the money. What are we going to do in 15 years, if we have no grant, no endowment to care for the animals? Washoe's the oldest and she could live another 25 years; chimps can live 50 years or so in captivity. What will happen when we're 75 or 89? . . . What are we going to do when we're not strong enough to care for her and we still have no money?"

When the Foutses first moved to Ellensburg, a representative of Progressive Animal Welfare Society, a Seattle-area animal rights group, contacted them. Roger Fouts was told angrily on the phone that the society did not approve of caged chimpanzees. "I said we don't like it either," Fouts recalls. "What we've got is awful compared to Africa. I invited them to come take a look and see what we were trying to do." Although the Seattle group has organized some tough demonstrations against researchers in Washington state, it has never singled out Fouts as a target. No one has ever protested his work in Ellensburg.

It was enough, in the beginning, simply not to be on the wrong side of the animal welfare movement. He wasn't ready to go farther, to become actively involved in crusading for chimpanzees. That came later, as the 1980s unfurled and as the animal rights movement grew in power. Animal advocates' rising anger against laboratory abuses, against the refusal of the scientific community to listen, in some ways paralleled his own. And it was the crusaders who came calling for him—with interest and with respect—rather than members of his own profession.

In 1986, an underground animal rights group called True Friends broke into SEMA, Inc., an NIH-funded laboratory in Rockville, Maryland. Scientists there did AIDS research. They had close to 500 apes and monkeys in the laboratory. When True Friends broke in— apparently tipped off by an unhappy employee—they brought video cameras with them. In the classic manner of such break-ins, they were interested in publicizing laboratory conditions. For their purposes, SEMA was perfect.

The AIDS-infected animals were boxed up alone. That meant each in a metal cube, with one small window. These "isolettes" were 40 inches high, 31 inches deep, 26 inches wide. Inside the boxes, animals were rocking back and forth in the blind, ceaseless motion of the mentally ill, of children who are emotionally starved. The resulting videotape was released through People for the Ethical Treatment of Animals, among the largest and most aggressive animal advocacy groups in the country. PETA sent the tape to television stations and newspapers. One copy was also sent to Jane Goodall in England.

Goodall's specialty was wild chimpanzees. But her long relationship with them had made her feel—as Fouts did—responsible. She felt, also, an obligation to speak out in their behalf. She was already in contact with Roger Fouts, and after seeing the tape, she asked him to visit SEMA with her. They were escorted by a top-level NIH administrator. Fouts, who knew caged chimpanzees better than Goodall, was still shocked: "It was a nightmare there. They had these chimps in metal boxes. One wasn't even rocking anymore. She was just lying on the floor of her cage. When we walked in, the chimp lifted her head and looked up. It was like those children you see in Somalia, that blank look. They're not there. And the vet said, 'See, she's not screaming,' and he told the tech to take her out. 'See, she's just fine.' They were holding her like she was a typewriter and she was just lying there."

Afterward, driving away, Fouts remembers, the government officials began cheerfully remarking that they must be reassured by what they had seen. Clearly, the facility met NIH standards. Fouts and Goodall sat silently in the back of the car. He looked over at her. She was crying, tears dripping off her chin. The experience convinced Fouts that NIH was indifferent to the animals. It left him thinking that, if he was afraid to speak out, the chimpanzees would suffer for it. (Eventually, public pressure forced SEMA to make some concessions. Reorganized, and under a different name, it continues the AIDS research on chimps that are still isolated, but now in big, see-through plexiglass units decorated with jungle scenes.)

It was in the late 1980s that people began to realize chimpanzees were in desperate trouble in the wild. The population in Africa had dropped from 5 million to under 200,000. The U.S. Fish and Wildlife Service, which had listed them as threatened, announced that it wanted to declare them endangered. The agency received more than 40,000 letters of support and six of opposition. Five of those six were from biomedical researchers; two came directly from NIH. The researchers were afraid that the endangered listing would interfere with use of captive chimpanzees. Researchers didn't want to import chimpanzees; they had no objection to classifying wild ones as endangered. They wanted laboratory chimps, though, to remain classified as "threatened," one step away from the more dire "endangered." The more serious listing would make it very difficult to use the lab animals, requiring a massive amount of paperwork and justification.

NIH's position carried an implied threat of sorts, at least to Fouts. The agency was breeding chimpanzees for medical use; if that effort became too difficult, under the changed listing, it might just have to kill the program. In a letter to the wildlife agency, NIH administrators suggested that it would be a terrible loss if the biggest chimpanzee breeding program in the United States, involving more than 1,200 animals, had to fold because of a bureaucratic classification. That it would be a blow against conservation, instead of a protection measure.

Fouts was furious. How could NIH claim to support conservation when its chimpanzees were being used in AIDS work, which ruined them for breeding purposes? He believed that the split ruling would send a clear signal that the United States wasn't all that worried about chimpanzees anyway. Not only did he complain directly to NIH, he took a more radical step. He joined with PETA and five other activists groups in a formal protest to the U.S. Fish and Wildlife Service. He wrote such a blistering critique of the NIH program to members of Congress that the agency was forced to issue a six-page letter of rebuttal, described by one extremely irritated administrator as the longest letter sent from the agency's animal research division to any legislator. Apparently, the rebuttal was persuasive. Fouts's efforts were in vain. In 1992, the service split the classification.

Roger Fouts was reminded where the power was. He also began to believe that NIH was shutting people like him out. He suspected that the agency fought the endangered classification because it would mean filing detailed reports on every animal used. All those reports would be public record, available to him, available to PETA. "Then researchers using chimps would be under greater scrutiny," Fouts says. "The whole mentality of NIH is to do it in secret. They're God, the keeper of the gates. You have to do it their way or you don't get in. And I am embarrassed sometimes to be considered a member of the priesthood."

One thing was clear though, he was no longer the forgotten scientist in backwoods Washington. He had become—for better or worse—a recognized voice in a growing cause. The 1980s had been a heady period for animal activists. They were engaging national attention and they were making people think about when—or if—animals should be used in research. They were sometimes viciously combative. During the late 1980s, the FBI reported more than 50 cases a year of lab vandalism and personal attacks on researchers. By the end of the decade, membership in animal advocacy groups had reached 10 million. Those were impressive numbers, especially in a political sense. The movement not only caught the imagination of the public, but the attention of the country's legislators as well.

In 1985, Congress decided to strengthen the national animal welfare act. The new law, which Christine Stevens had lobbied for fiercely, included a special section for captive primates, chimpanzees, baboons, and other monkeys. It insisted that researchers had to care for the animals' "psychological well-being." The U.S. Department of Agriculture, the agency responsible for inspecting laboratories, was left to figure out exactly what that meant. And USDA was baffled. Was it responsible for animal happiness or merely contentment? Its officials decided to bring together a panel of the primate experts.

The committee was chaired by Michael Keeling, who heads one of the NIH-backed chimpanzee breeding and research facilities in Bastrop, Texas. Most of the panelists were from big institutions. Among them were Roy Henrickson, chief of lab animal care at the University of California, Berkeley, and Frans de Waal, famed for his work in the social behavior of chimpanzees, from the Yerkes Regional Primate Research Center in Atlanta. But USDA also asked Roger Fouts, guardian of five chimpanzees, to come. It invited Jane Goodall, too. She sent, instead, a letter with Fouts, pleading for better care of captive animals.

Goodall's letter got things off to a wrong start. As some of the other panelists recalled, "Fouts came in, waving the letter, and obviously thinking that since he was Jane Goodall's representative, we would do whatever he wanted." Fouts's remembrance was that he was treated as an outsider from the beginning: "I found out very quickly that I was in a minority," he says. "There were mainly biomedical people from huge labs there." He was just a psychologist from a small university, representing a lab with five animals. He had a cause though. He wanted laboratories to stop boxing up animals. He wanted bigger cages. He had SEMA at the back of his mind.

Scientists tend to be stingy about cage size. NIH had revised its cage standards in the early 1980s, but only after a panel found that some monkeys were packed into enclosures so small that they couldn't turn around. Even so, the panel was bitterly criticized by researchers who were forced to buy new cages. They complained that there wasn't enough scientific justification for the change. The minimum cage now for lab monkeys, who can weigh more than 20 pounds, is 3 feet by 2 feet by 32 inches high. It's a box, and research lobbyists have persuaded USDA anything bigger would be overindulgent.

Animal advocates emphasize the cages, in part, because they can be measured. It's a standard by which to judge whether a laboratory is doing a good job. If you don't trust laboratories—and animal activists never have—then you want the measurements down in writing. The big laboratories, though, don't want to report to people like Christine Stevens. They don't want to be told what to do by people they hate. They suspect that the push for bigger cages is simply to raise costs and drive them out of business. To the annoyance of some of researchers on the panel, Fouts didn't seem to care about that.

"The one thing I was pushing, the old regulations said you could put an adult chimp in a five by five by seven cage and feed and water him once a day. When cage size came up, I assumed people would agree that was inadequate. Instead, they kept talking about cost. They would say, I've got 1,000 cages, it's too expensive to change them. I said, come on, it's nuts. It's insane to have a law that says you can legally put a 600-pound gorilla in a cage that size. I guess they thought I was naive. I made the motion, no one seconded. Keeling said, 'It's obvious no one else thinks that's a good idea, Roger.' I said, 'Well, at least give them 50 square feet.' And no one would second that either."

When the final USDA rules came out, there was nothing in them that required bigger cages. In fact, they actually gave scientists a means to use smaller cages, if they could justify it to USDA. If Fouts had doubted that he was an outsider before, the panel convinced him he was. The best response he had gotten from his colleagues was sympathy.

"Chimpanzees are incredibly special," Henrickson says. "They're so smart. I've always been glad that I didn't have to work with chimpanzees in biomedical experiments. I wouldn't like to make that choice. And if I was isolated up in Washington, working with chimpanzees every day, I might have become Roger Fouts too."

Fouts's old friend Christine Stevens, from the Animal Welfare Institute in Washington, D.C., called immediately. Like him, she was appalled by the USDA regulations. Not only did they back away from cage-size requirements for monkeys, but for other animals as well. They didn't even promise increased exercise for lab dogs, one of the top goals of the advocacy groups. And in the case of psychological well-being of primates, they allowed each laboratory to come up with its own plan, which would not be available to the public. Essentially, Stevens said, the regulations gave it back to the laboratories and said, whatever you think is best.

There was no doubt in Stevens's mind about what to do next. She was going to sue. She wanted the lawsuit to stand as an accusation against three agencies: USDA, as administrator of the law; the U.S. Department of Health and Human Services, the parent agency of NIH, which had lobbied so hard for meaningless regulations; and the White House Office of Management and Budget, which had continually introduced cost-cutting factors into the regulations. She'd already lined up a lawyer, Valerie Stanley of the Washington, D.C. firm of Galvin, Stanley and Hazard. She and Stanley had agreed that they needed a scientist to stand as a plaintiff in the suit, a member of the research community who found the regulations to be harmful. Roger Fouts was their first choice.

"It crossed my mind, you know, to just back off NIH," Fouts says. "I thought, I knew, there was a high probability of never getting another dime if I joined with the lawsuit. But you have to be able to sleep at night. In the end, I thought, I have to speak out. How am going to live with myself if I don't?"

The first name of an individual plaintiff on the lawsuit is that of Roger Fouts.

The lawsuit went to U.S. District Judge Charles Richey in Washington, D.C. To the dismay of biomedical researchers, Richey agreed with Fouts and Stevens and their friends. Richey issued his ruling on February 25, 1993. He called the USDA's refusal to upgrade cage sizes "arbitrary and capricious." The judge himself hinted that he suspected biomedical researchers of having written the rules to their own liking. He insisted that larger cages should be available for captive animals—and soon. Judge Richey told USDA to try again, and this time not to rely quite so much on the "good faith" of the people it was trying to regulate. In November 1993, the U.S. Department of Justice, which represents USDA, told Richey that the agency will appeal every point of his ruling. In fact, an appeal was inevitable even before that. Two powerful scientific organizations, the National Association for Biomedical Research and the Association of American Medical Colleges, had announced their own plans to fight the ruling. Researchers, Stevens says, despairingly, will do almost anything to avoid increasing cage size for laboratory animals.

And Roger Fouts? Has he, in these battles, accomplished anything for research chimpanzees? If he had stayed in the system, could he have influenced it better from within? Undoubtedly, if he had been willing to use animals, without sentiment, he might be doing more innovative science, might have gained a professional reputation that would have given him more clout. Has he sacrificed his career only to end up an annoyance at the edges of the issue? They are the kind of questions that come up in a sleepless night. When asked, he shakes his head in a kind of slow resignation. "If I focused only on actually achieving change, I'd have to quit," he says. "It doesn't happen that way. I'm doing this, partly, so I can face myself in the mirror in the morning." Despite his money problems in Ellensburg, he's provided well for his own animals, for now. He's overcome the somewhat grudging attitude of his own university, moving his chimpanzees into a new building in May 1993. The new home was paid for by state and private funds, raised through tireless fund-raising by his foundation. The five animals share 7,000 square feet of living space, with both indoor and outdoor runs.

Ellensburg has not been the haven he once hoped, though. The Cascades, for all their forbidding majesty, are no defense against his critics, against bad news from other research centers, least of all against his fears. He still thinks, too much maybe, about the chimpanzees in laboratories on the other side. He's begun to think beyond them, as well. Roger Fouts, clearly across the line, now wonders if working for chimpanzees is enough:

"If we can ever get people to accept great apes and chimps, then that will extend to other primates, baboons, rhesus monkeys. It's just a matter of time until, eventually, that turns into a reverence for all life. We have to take chimps first because they're so close to us. Chimpanzees are the first battleground."

TWO

Of Street Toughs and Target Practice

IF HE HADN'T spent his career with chimpanzees—if he had started by studying rats, or pigs, or even your basic monkey—you have to wonder if Roger Fouts would have taken quite so many risks, if he would have cared so much. Chimpanzees with their mastery of sign language, their intelligence, their close kinship with humans, seem to bring out protective tendencies in many scientists. That hasn't been so true with other animals, even other primates. Even today, there is little energy spent worrying about the possible mental anguish of mice and rats in laboratories. Duane Rumbaugh, a long-time expert in animal intelligence, remarks that, at least in the past, researchers often treated monkeys as if they had the mentality of a slice of baloney.

In the research world, people are only just beginning to appreciate how smart monkeys are. Rumbaugh himself helped show that they could perform tasks once assumed too difficult for anyone but a human, or maybe a chimpanzee. Such studies begin another round of boundary blurring. The ape-language studies helped erode the border between people and chimpanzees. Now the "lesser" primates are coming over the wall, moving into that charmed circle of those creatures deserving of special respect, smart enough to reason and negotiate their way through life.

Rumbaugh's own appreciation of monkey intelligence began almost by accident. He couldn't have forseen it at the time. But he began moving toward a major discovery of monkey intelligence during an argument over chimpanzees. That discussion took place in the unlikely setting of Lenox Square Mall, an upscale, sprawling shopping complex in north Atlanta. It was November 1984. The mall's indoor trees were spangled with tiny, white lights, advertising the approach of Christmas. Sue Savage-Rumbaugh was thinking video games, thinking about buying a new computer. Duane Rumbaugh was against it. The husband and wife stood arguing, in the twinkling shelter of a fichus, ignoring the jostling shoppers around them.

The shoppers ignored them too. How many hundreds of couples have stood in malls, arguing about expensive toys during the holiday season? Savage-Rumbaugh didn't want the computer for a toy, though. Her mind wasn't on the holidays at all; she wasn't planning a gift for her son. She wanted to find out if her two research chimpanzees, Sherman and Austin, could play computer games, that pastime of modern human adolescence. She wondered if they could hold a joystick in their hairy, dexterous hands and use it to shoot down targets.

"Don't do it," counseled Rumbaugh.

He was already one of the country's most renowned researchers in primate intelligence. Now Regent's Professor of Psychology at Georgia State University, Rumbaugh, 64, had spent years pursuing questions of animal intelligence. He'd looked at rats; chickens; tiny, graceful squirrel monkeys; even one belligerent gorilla that had stymied the experiment by smashing the equipment. As for chimpanzees, he knew how smart they were. He still remembered how astonished his group had been when their first chimpanzee, Lana, began typing out demands for bananas and affection. The Lana studies, in fact, remained the foundation for all work at their laboratory. The keyboards were computerized, but they were hardly Nintendo.

Despite their image of noisy, mindless amusement, computer games require complex thinking. You have to connect totally unconnected things—the rubbery joystick in your hand, the dancing images on the computer screen. You have to recognize that what you do with one affects another. You have to analyze motion—the blast of a rocket across the screen—and respond to it accurately, aiming and firing your "guns." It was one thing to ask a chimpanzee to tap a key and receive food and drink; shooting down targets on a screen was a major leap. Too wide a leap, he thought.

Savage-Rumbaugh was also a leader in the field. Like Fouts, she had worked at the University of Oklahoma's well-known primate research institute before moving to Georgia. She and Duane Rumbaugh, now married more than 16 years, ran the Language Research Center at Georgia State University, devoted to a singular search for the limits of primate intelligence. Their work had received complimentary coverage from journalists around the world, from Japan, from London's BBC network, from the *National Geographic*. And in Rumbaugh's opinion, their prominence made the computer game idea more risky. As he knew from experience, there were always critics poised to take shots at any study of animal intelligence, anything that seemed to chip at humans-only capabilities.

"If we buy the computer and we fail," he told her. "Everyone will think we've gone off the edge." He knew, really, that he wasn't going to change his wife's mind. Like him, she had a sure, tough faith in her own instincts. It was one of the reasons they worked well together. In the end, it didn't surprise him that the machine was in the car trunk as they drove out of the mall parking lot. He argued anyway. They squabbled all the way back to their laboratory, through the heavy traffic on Interstate 85 south and the dense, piney woods of the laboratory grounds.

The Rumbaugh's primate laboratory looks like a cluster of boxes—beige, flat-roofed rectangles—tumbled into the trees. Inside, there are small offices and glass-walled laboratories, opening up to cages that look down the hillsides into the woods. On sunlit days, the chimpanzees move into the outdoor cages, playing together, out-staring the squirrels as they rustle warily by.

Savage-Rumbaugh spent several weeks in her own office there, designing a simple computer program—using a joystick to move a cursor to a box-like target. When it was ready, she put the monitor into a laboratory. Then she brought in Sherman and Austin. Rumbaugh, reluctant to watch, braced himself. Savage-Rumbaugh sat in a chair and began playing the game, demonstrating it to the chimpanzees. Sherman and Austin watched, silently, intently. Then, Sherman advanced and gave her a shove, trying to get Savage-Rumbaugh out of the chair, to take her place. He climbed up next to her and started playing the game. Then Austin started shoving at Sherman.

"Within ten minutes, they'd literally pushed her out of the way." Rumbaugh says. "In the beginning, she and Sherman competed quite a bit, and he did extremely well. The first time he beat her, she pretended to be very disappointed. And after that, there was a suggestion that he was deliberately slowing down his responses sometimes, letting her win."

Rumbaugh loves to tell that story. It always makes him laugh.

There's just something about being so completely, outrageously wrong that really appeals to him. And, of course, he ended up having the last laugh.

About two years later, in 1986, the National Aeronautics and Space Administration approached him with an unusual assignment. The agency planned to conduct a major life sciences mission in 1996, turning the space shuttle into an orbiting biology laboratory. One of the experiments would require two monkeys to be on board.

Among all the questions that space scientists had about life outside the familiar tug of Earth's gravity, one was fundamental. Could mammals, primates, humans survive there? You could forget colonizing the moon if the bodies of the colonists started slowly dissolving into gelatin. It was a real worry; there were hints of it in every space mission. Without gravity's drag, astronauts began to suffer from wasting muscles, thinning in their bones, even in the short shuttle missions.

The longer the stay, the worse the effects. In the old Soviet space program, cosmonauts had stayed aloft for more than a year, floating between the curved walls of the station Mir (Russian for Peace). Despite carefully designed exercise programs, some of the men could hardly stand on return to Earth. Some had never fully regained their bone mass. Until that problem was solved, human society appeared—except for short flights—grounded.

Beyond that dilemma, researchers like Rodney Ballard, the chief life scientist at NASA's Ames Research Center in California, thought if they could figure out what made bones weak in space they might be able to make them stronger on the ground. Ballard thought if he could decipher the process of bone disintegration, he might gain insight into the reverse effect—what holds bones together. Perhaps that knowledge would allow them to speed the healing of broken bones, block the failure of aging hip joints.

His agency, in collaboration with French space scientists, had decided to put two monkeys on the shuttle. The animals' bone mass would be tested before and after flight. They would have a consistent diet, a consistent routine. Scientists on Earth would get daily measurements on their body temperature and heart rate, from sensors relaying data to receivers. You couldn't study the astronauts like that. People didn't follow the same routines; some of them wouldn't put up with the tests; they wouldn't eat the same food. One study had been loused up by an astronaut who insisted on dumping Texas hot sauce all over every meal. Another study of proper diet had been lost when the entire crew was buffeted by space sickness and refused to eat at all for several days.

So, NASA scientists settled on a species of monkey known as rhesus macaque (pronounced ma-KACK). There are only about 20 species of monkeys widely used in biomedical research. Rhesus macaques head the list, followed by their close cousins, the crab-eating macaques of Indonesia, pigtail macaques and stumptail macaques. The most commonly used other species are squirrel monkeys, small South American tree climbers, and big African baboons. There are others—marmosets, titi monkeys, owl monkeys, African green monkeys, mangabeys, tamarins—but they are minor players.

Rhesus macaques are tough monkeys. They held the promise of being good space travelers. Researchers at Ames had sent macaques whirling around on a centrifuge. That dizzying ride made squirrel monkeys acutely sick. The macaques were ready to saunter away. They were considered among the most adaptable monkeys, not in terms of temperament, but in ability to survive. Natives of the forests of Asia—India, Bangladesh, China—they had learned to live in farms, temples, city streets, any available backyard as the forests were slashed down around them. Biologists class them as among the planet's most indestructible animals. The uncomplimentary term is "weed species," those capable of making a home anywhere. Other weeds: rats, pigeons, raccoons, coyotes, sucker fish, human beings. Surely, reasoned the NASA scientists, the monkeys could adjust to a short ride on a space shuttle.

There are monkeys—fluffy, dark titi monkeys from South America who sit close together, tails entwined like lovers—who make you want to cuddle them. Rhesus macaques aren't like that. They look like the street toughs of the monkey world. Wiry, golden-gray fur frames a squared, smooth face, dominated by a long snout and a pair of close-set eyes, coffee-dark, intent and always watchful. Rhesus macaques are large monkeys. Big males reach a muscular 30-plus pounds. But it's their teeth that stand out. The canine teeth of a rhesus macaque are dagger-like. The animals are quick to draw back their lips—so casual it can look like a yawn—and display their weaponry. In American research labs, they've gained a reputation for playing rough, fighting rougher. If they can't bite, they'll scratch. They are as fierce with human handlers as with other monkeys. They've been known to yank on each others' tails so viciously that they rip the skin completely off. Vets call it a "degloving injury."

There are scientists who like them very much, but there's a certain respectful distance in their descriptions. They speak of fierce independence and pride. Rhesus macaques can be born and die in captivity and never seem domesticated, only stubbornly wild. Yet, they share one critical quality with humans. They are intensely social, frustrated by boredom, desperate in loneliness. For all their apparent toughness, they are animals that survive by family bonds. They live in a society made of social classes, with its own aristocracy and peasantry. They care for each other, grooming, cleaning each other's wounds, babysitting children. When danger threatens macaques, they do not pull apart into independent fighting units. They huddle together. Without companionship, captive monkeys have been known to fall apart: rocking themselves into trances, hunching in corners, pulling their golden brown hair out in clumps, sometimes picking open their own skin.

The space monkey experiment, called The Rhesus Project, calls for a two-week mission, with the monkeys in isolation. Each of them is to be completely closed in a metal and plexiglass box so that they could see each other but not touch. Neat, crisp technology, but still NASA's Ballard feared for the monkeys' ability to endure the separation. The last thing NASA scientists wanted was to have their astronauts step triumphantly out of the space shuttle followed by technicians carrying two bloody and neurotic monkeys. NASA wasn't trying to study stress. Rodney Ballard was determined not to. He wasn't going to cripple a good experiment through careless handling of animals. Colleagues thought he was becoming obsessive; one impatient NASA scientist suggested he simply throw a rag doll in with the monkeys. Ballard's instincts told him something else, that solving their monkey problems might lead them to some very good science. He started asking around. And he started hearing the same answer, ask Duane Rumbaugh.

So Rumbaugh got the call: What to do with a monkey in a box bound for orbit? He puzzled over it; he knew as well as anyone how quickly an isolated rhesus macaque could collapse. Still, this was NASA, with its love of technological wizardry. Suddenly, stubbornly, the memory of Sherman, Austin, and the video games sprung up in his mind. You know, he said to his wife, I think I'm going to try the rhesus monkeys on the computer. Like his wife, he wasn't really thinking of entertainment. For chimpanzees and monkeys, shooting down a video target might resemble a game, but it was actually a planned task, a challenge. "Don't do it," she said, beginning what was almost the mirror-image of their argument at the mall. Her ammunition was better, though. The classic scientific literature, stacked to the ceiling in the Rumbaugh's book-crammed lab and home, said that rhesus macaques would never be able to figure out a joystick. The literature said, if they couldn't touch the screen, they couldn't understand it. Still, Rumbaugh couldn't help remembering how wrong he had been about Sherman and Austin. Somehow that memory kept nagging at him, urging him to gamble.

"I knew I'd have to invent an approach," he said. "And I thought, well, I didn't think the chimps could use the joystick and now I don't think the rhesus can use the joystick. So I think I'll try it out. Sue said don't do it, but I modeled myself after her and didn't listen. I told NASA that there was a high risk and it probably wouldn't work, but that if it did, we'd be limited only by our imaginations.

"We had two rhesus macaques already here, Abel and Baker. We showed them the joysticks and it was just like falling off a log. We started in March, and by June, I was giving a presentation to a joint NASA-French symposium. It just blew everyone's mind."

He grins, thinking about it, blue eyes bright with amusement: "Thank God we didn't let the rhesus into the library to read the literature."

Not only did the rhesus macaques take to the computer games, but they were good, really good. Holding the joystick in their small hands—a young macaque's palm is perhaps half-an-inch across—they learned to operate the equipment with precision: to chase, to capture, to pursue moving targets, to shoot bullets of light and hit targets. It wasn't only the joy of the chase. They received food rewards—banana-flavored monkey pellets—for each destroyed target, each correctly aimed light bullet.

But, as Rumbaugh and research associate David Washburn rapidly discovered, the monkeys liked the work, the challenge, even more than the food. When the scientists left Abel and Baker in front of a computer, they would return to find the monkeys absorbed in the work, holding a monkey biscuit absently in one hand while plying the joystick with the other.

More surprises followed. In the chimpanzee laboratory, the scientists were exploring the chimps' ability to understand the relationship of numbers: one is less than two, two is less than three, that sort of thing. Again, they were using computer programs, asking the animals to use a joystick to guide a cursor to the correct answer. As they carefully designed the chimpanzee experiment, Rumbaugh began trying to work out the bugs in their software. For pure variety, he decided to run the program for Abel and Baker, letting them watch something different, working out the kinks in the program. To his astonishment, they began, slowly, to figure out the relationships between numbers.

"If you view monkeys as operating on basic stimulus and response—and a lot of people still do that—then that's all you're going to get out of them," Rumbaugh says. "If you don't even look for intelligence, there's no way you're going to see it. What happened here was that we were testing out our software for bugs, we were going to use it for the chimps. And the monkeys just picked it up." The animals had to choose from two numbers that appeared on the screen, between zero and five. Zeroes appeared often, in every other pair. For each number chosen, the computer would release that number of food pellets. None for zero; five pellets for five. The monkeys came rapidly to prefer five to one. And no matter how many times zero flashed, they didn't chose it. It was a loser and they knew it.

Okay, but that was a pretty simple relationship. So, Rumbaugh and Washburn decided to push the rhesus macaques a little harder. They put in a new software program, pairing numbers from zero to nine. They would show the monkeys some of the pairs, say three and six, four and eight. The higher number always carried a higher food reward. They didn't show them every possible pair though. They might not flash three and eight on the screen. They'd wait until they were sure the monkeys were comfortable with the other groups, then they would show the strange numbers. Choose the highest: three or eight? Almost always, the macaques chose eight. The researchers began to wonder if their ability to compare was limited to only two numbers. So, they flashed five numbers up on a screen. On the first trial Abel chose the highest number of five choices more than 72 percent of the time.

Back at NASA's Ames Research Center, Ballard and his colleagues were stunned. Rumbaugh's results seemed amazing. They found themselves strolling out to look at their colony of macaques. They were greeted with the familiar, unfriendly stares. There was a feeling that maybe Rumbaugh had stumbled onto the two most brilliant rhesus macaques on the planet. Maybe Abel and Baker were monkeys of Guinness Book genius. There were about 75 macaques in their research colony. NASA scientists decided to put 30 of them into training for the shuttle mission.

Ballard expected that half of them would drop out, fumbling with the joystick, baffled by the moving targets. The rhesus macaques turned out to be better at the computer tasks than some of their trainers. Worried that the monkeys would go hungry, not earning enough food through their game-planning, the NASA scientists had provided them with open boxes of food. The monkeys ignored the boxes as food rewards piled up around them. Finally, Ballard quit waiting for dropouts. He began almost arbitrarily reducing the primate computer corps, narrowing his way to the two monkeys that would ride the shuttle and two backups.

"They're all really bright. It comes down to subjective judgment at the end," he says. "There's always four or five animals that everyone just really likes."

For Rumbaugh, the work was a revelation. How badly had people underestimated rhesus macaques? How complex were their thought processes? He began drawing up even more ambitious plans for Abel and Baker. It was possible, he thought, that like chimpanzees, they could learn a language system, communicate with their human trainers. He had to wonder: Will macaques one day tap out a simple sentence, asking for a tickle, the way chimpanzees do?

If rhesus macaques can do that, what about the other thousands of monkeys now used in U.S. research laboratories? What about crab-eating macaques, stumptail macaques, pigtail macaques, bonnet macaques? What about baboons, with their long noses and clever hands? African greens, with their talkative family groups, or friendly, dark-eyed sooty mangabeys. Not to mention the South American monkeys: squirrel monkeys, owl monkeys, gentle titi monkeys, cotton top tamarins, tiny marmosets, with their rounded noses and tufted ears? What about sable-brown capuchins? Capuchins, like chimpanzees, share food with each other, trade treats. They have the largest brains of any monkey in proportion to body weight.

The intelligence of chimpanzees has brought them to the forefront of the battle over primate use in medical research. Still, in American laboratories, they're barely visible compared to other animals. What if you made the same intelligence connection for macaques, the most widely used research monkey? It's one thing to be highly selective in using chimpanzees. If scientists became reluctant to use macaques in research, that would come very close to emptying out the primate labs. Raising the issue is a Pandora's box. And Rumbaugh knows it.

"Knowledge is risky in that you learn things that sometimes burden you," Rumbaugh says. "But the first basic commitment is to learn the truth. Then you deal with the consequences, ethically and morally. It's not Alice in Wonderland here. We have some control over what happens." We also have, he thinks, an obligation.

"Primates are a truly remarkable life form. We need to recognize their psychology. They are not to be responded to as a rock to be bulldozed aside. These are sentient animals. They have an intelligence that is quite similar to our own."

In its own way, that is a revolutionary statement. Supporters of research have argued the opposite for hundreds of years. The philosophy dates back to the year 1637. That was the year the French philosopher René Descartes, famed for his passionate belief in the intellectual powers of man—"I think, therefore I am"—made his stand on the great divide between humans and other species. Descartes was unhesitating. He dismissed everything on the nonhuman side of the divide as walking machinery. No mind, no soul and, therefore, no value except, of course, as tools for the use of humankind.

It's hard to debate a dead man, but Rumbaugh finds himself trying to. He's spent too many years being surprised by animal intelligence not to be angry at such sweeping assumptions. He believes in doing research with animals. That's not the issue for him. The issue is doing it blindly, which to him means badly: "The Cartesian principle is one of the saddest things to be put forth as a certainty. It was completely nonscientific, the idea that without a soul you couldn't experience pain. Pain was God's punishment. Animals, who had no soul, might present the appearance of pain, but they didn't really feel it. So, from Descartes's point of view, the behavior of animals was autonomous, the mindless turning of wheels."

As he talks, Rumbaugh is driving rapidly on the rutted dirt trails that link the laboratory with his house. Five years ago, the Rumbaughs bought land near their laboratory, built a modern, peak-roofed house, framed with wood, walled with glass. Standing in their living room, barring the furniture, is like standing in the woods. Rumbaugh loves the woods and knows them blind. The old brown pickup plunges down hills, over rocks, corners across piles of fallen leaves, slides around tree trunks. But at the house, he stops to pick up one of his pets, an aging sheepdog, its golden fur fading to cream. He drives more gently with the dog in back. His passenger quits holding her breath.

He flexes his hands on the steering wheel, "The more enlightened perspective is that these animals are related to us. The roots of what made us what we are can be found in them. They don't just appear to suffer," he says, his voice quiet as the trees blur across the glass. "They do."

IF YOU ACCEPT THAT, that animals suffer, then you accept that knowledge, based on animal research, has a price in pain. The question of intelligence is less straightforward. As smart as a chimpanzee or a monkey is, can it comprehend that it is being used by another species? Does it recognize the experiment that can lead to its own death? No one has pushed the intelligence question that far yet— and from the research side, it is not a question that everyone wants to see answered.

The use of animals in research has been with us for nearly 2,000 years, certainly throughout the era of Christianity. Back in the second century A.D., the Roman physician Galen was dissecting pigs and apes, trying to understand the functions of the body. The law barred him from cutting apart human cadavers.

Primate research as we know it, though, is really a modern pursuit, belonging to the last century. If we are going to question it, to consider whether it should be reexamined and reshaped by questions of pain and intelligence, we must first consider its more recent roots. And we must consider its path, what drove it by leaps and bounds, by stumbles and pure luck. The accounting must include the knowledge gained, the human lives saved, or just eased, through discoveries traceable to the use of monkeys in medical research.

Science historians credit a British physician, David Ferrier, as the first to look at the primate-human relationship in a systematic way. Ferrier was a physician at King's College Hospital in London. In 1876, he published a study of brain functions that made direct comparisons between human and monkey brains. And apparently overnight, there was collective fascination by scientists. Soon, others joined Ferrier and began opening the brains of bonnet macaques and orangutans, looking for parallels.

The work of Charles Darwin, suggesting a close genetic relationship between man and monkey, lent a new force to the doctors' observations. There was a new sense of connection. By 1917, one British scientist, Sir Charles Sherrington, had dissected the brains of twenty-two chimpanzees, three orangutans, and three gorillas. He was looking for the similarities to humans.

Researchers also were realizing something else. It was slowly becoming obvious that, whatever the relationship between human and nonhuman primates, they were susceptible to many of the same diseases.

In 1908, Viennese researchers Karl Landsteiner and Erwin Poppert conducted an experiment on monkeys, brilliant in its simplicity. They removed part of the spinal cord of a boy who had died of polio. They ground it up, filtered it, and injected it directly into the spines of two monkeys. One became paralyzed. Both died. Their spinal cords showed the same painful damage as did those of humans infected by polio. Clearly, then, monkeys could be a remarkable model of human illness. Other scientists rapidly caught on. In 1911, monkeys were found susceptible to measles. In 1914, mumps. In 1928, yellow fever.

There was none of the genetic linkage known today. No one had yet identified the twisting structure of DNA, deoxyribonucleic acid, on which is stamped the genetic instructions for life. No one had studied the patterns of genes in humans and other species. Scientists of the time had no way of knowing that the DNA of humans and rhesus macaques is about 92 percent identical. But the circumstantial evidence was compelling: even without the underlying biology, any doctor could watch a monkey fall ill and die of a human disease. The circumstantial evidence was undeniable.

Like any good scientist, Landsteiner wanted to take that idea and run with it. But he had a problem: He couldn't get the animals. Bringing wild primates out of the wild was prohibitively expensive. It took weeks, months by train and boat. Few animals survived the trip. Landsteiner's superiors told him to forget monkeys and to use rats, cheap and easy to get. Fascinated though, Landsteiner continued working with the few macaques he could get. His most famous medical work had been the discovery of blood types in humans; it won him a 1930 Nobel Prize in medicine. He decided to pursue the same kind of question in primates, breaking apart the chemistry of blood. In 1940, Landsteiner and his closest pupil, Alexander S. Wiener, announced the discovery of the "Rh" factor in blood—Rh for Rhesus, as in rhesus macaque.

The Rh factor refers to a cluster of highly reactive proteins on the surface of red blood cells. Most people are Rh positive; the proteins are there. A few lack the proteins; Rh negative. A little difference, but it could have a stunning effect. If a mother is Rh-negative and her offspring positive, her body will accept the first child. But her body will, as a result of exposure to that first, build up a hostile defense against the next Rh-positive child. Her immune system will rip into the fetus' alien red blood cells, jamming them together. Before the discovery of Rh factor—shared by humans and rhesus macaques—several thousand children every year were born brain-damaged or dead as result of such incompatibility. The majority of children in state mental hospitals were the result of Rh complications. With Landsteiner and Wiener's discovery, researchers were able to develop a vaccine that blocked the vicious immune response.

Wiener became so passionate about the value of primates in research that he eventually helped start the Laboratory for Experimental Medicine and Surgery in Primates (LEMSIP) in New York. Jan Moor-Jankowski, the laboratory's current director, credits two advances in science and technology for making both that facility and modern primate research possible.

The first was the rise of the airline industry, especially after World War II. With that kind of rapid transit, it became possible to rush animals out of Asia and Africa, eliminating suffocating deaths in the holds of ships. The second advance was drugs, specifically sedatives. Wild monkeys were not domestic kittens. They could not be stroked into calm. They bit and they were dangerous. Rhesus macaques were the first lab monkey used with any regularity and they rapidly earned their reputation for intractable fierceness. A monkey bite was more than just an ugly cut, too. Monkey saliva could be rich with infectious organisms. In the same way that human viruses could attack monkeys, monkey viruses could be genuine plagues in humans.

It was not until 1956 that Parke-Davis Pharmaceuticals in Detroit brought out an animal sedative and painkiller called Sernalyn, from the word "serenity." Up till then, primate researchers had been trying to control their animals by dosing them with drugs like phenobarbital, a tranquilizer so potent it often induced a fatal coma in monkeys. With a shot of Sernalyn, a monkey could be immobilized with almost no apparent side effects. It seemed incredibly safe. With most anesthetics, if you double the prescribed dose, you can kill the patient. Not Sernalyn. In monkeys, it took 26 times the prescribed dose to induce death.

If Sernalyn really had been the perfect sedative, of course, it would still be on the market. It turned out to be one of those compounds that prove the monkey model for humans less than ideal. The wonder sedative in monkeys turned out to be a perfect nightmare in humans. In the early 1960s, because of the success in animals, doctors began testing Sernalyn on their patients, as a possible human drug. The giveaway to what happened is Sernalyn's other names. The technical name is phencyclidine hydrochloride; the street name is PCP. The drug also goes by the names angel dust and rocket fuel. Narcotics enforcement officers could rattle off a half-dozen other slang terms.

There's no question that it tranquilizes monkeys and can do the same for people. It just doesn't stop there. In people it induces a warping of body image, a wandering sense of disconnection. In higher dose or with repeated use, it produces a terrifying blend of paranoia and blinding fury. Those side effects turned up as soon as clinical trials began, as soon as the drug was tested on human patients. The tests were stopped cold in 1965. The U.S. Food and Drug Administration abandoned consideration of Sernalyn as a legal drug. But by then it had a reputation for its mind-bending high, a reputation that has kept PCP in demand.

Actually, the word had spread early among primate researchers, too. UC-Berkeley's Roy Henrickson learned about Sernalyn firsthand, by accident. In the 1960s, Henrickson was chief vet at the California Regional Primate Research Center, one of the largest monkey research colonies in the country. Today, the center houses only monkeys. Then, there was also a small group of chimpanzees. Henrickson has a special affection for chimpanzees. He hated using a tranquilizer gun to immobilize them; he hated the way they looked back at him, with baffled pain. So one afternoon, when he was trying to dose his chimpanzees with Sernalyn, he decided to give it to them in Kool-Aid. The chimpanzees, who normally loved the sweet drink, loathed the bitter hint of the drug. They'd pick up the cups, take a drink, and spray the mixture across their cage. Henrickson, determined to drug them without forcibly injecting them, kept trying new formulas—a little less Sernalyn, a little extra fruit juice. He'd taste it, think maybe this time, it's tolerable. He'd pass it to the chimps and they'd spray it all over the cage again.

"You can guess the end of the story," he says wryly. "All at once the bottoms of my feet started to tingle. I completely tripped out. They had to carry me home and put me to bed. The chimpanzees were fine, though."

In the early 1970s, Parke-Davis replaced Sernalyn with a chemical cousin, Ketamine. No matter that Sernalyn worked well in monkeys, the pharmaceutical giant had no desire to continue making a drug increasingly famed as an illegal and dangerous hallucinogen. Ketamine, the only PCP analog that can be made legally, is only one-twentieth as potent a sedative. Ketamine is more of a depressant than PCP, shorter-acting, without the risky side effects. It is approved for use in people as well, as an anesthetic for high-risk patients, such as the very young or very old. It is also a preferred anesthetic in treating burn victims. There is no primate research facility in the country that doesn't use Ketamine on a daily basis.

First Sernalyn, now Ketamine—with the availability of such drugs, Moor-Jankowski says, researchers became able to handle monkeys in mass, to do research by the truckload instead of the handful. Also, they were learning how to keep transported monkeys alive. Physicians working in primate research, such as Moor-Jankowski himself, saw the symptoms of severe dehydration in crated animals. It was a collective shock of recognition, the newly arrived monkeys had the sunken eyes and crusted lips of dehydrated infants. The doctors began treating them like infants, using electrolyte fluids to dramatically reduce the death rate.

Suddenly, monkeys were available by the hundreds of thousands. Rhesus macaques were numerous then, so much so that they virtually could be shoveled out of the jungles of India. In the late 1950s, the United States imported more than 200,000 rhesus macaques a year. Through the early 1960s, the numbers held at well over 100,000 annually. Those numbers tell about more than access though. There was enormous demand for a reason. Beginning in the 1950s, the race for a polio vaccine was on. Scientists had discovered that they could grow the polio virus in monkey cell cultures. Everyone wanted rhesus macaques, the tool to developing a vaccine against a very dreaded disease.

In 1954, Jonas Salk showed that people could be protected against polio by inoculating them with dead polio virus, a killed-virus vaccine. His success led to a more intensive push for a vaccine made with a weakened live-virus, which would produce an even more powerful immune response. That vaccine, developed by Albert Sabin, was ready in 1958. Both men needed a free hand to deal in monkeys. Their work depended on it. The virus was grown in monkey kidneys and it was tested in monkeys. Researchers spent long weary hours peering into the animals' spinal cords to see if the polio viruses had been disarmed by the vaccine.

By all estimates, at least a million monkeys died in the race to halt polio; by some estimates, the toll reached five times that. The achievement has compelling human numbers as well. Before the vaccines, in the United States alone, 20,000 people a year were crippled or killed by polio viruses. In the early 1960s, when vaccine production was running smoothly, the numbers dropped to a few cases a year, cases suddenly so unusual that their appearance was startling.

You might believe, for the moment, that any human ill could be cured if enough monkeys went into the scientific pot. "Imagine," says Moor-Jankowski, "If Landsteiner had access to monkeys in 1908. We might have had the polio vaccine 20 years earlier. Tens of thousands of people would not have been paralyzed. If you weren't alive during the polio crisis—if you weren't there to stand on a street corner in Switzerland and see the people crippled by polio, just huddled on the sidewalk—then you can't know how bad it was until we had the vaccines."

And there was another story about monkey research, tragic rather than triumphant, equally compelling. In Europe, in 1959, the drug thalidomide came onto the market. It was a slick, powerful sedative, thought to be the perfect sleeping pill. No one thought twice about prescribing it to pregnant women, often desperate for rest. Many researchers of the time thought the human fetus was armored against chemical assault. No one worried then about mothers who smoked tobacco, drank alcohol, did a little cocaine. And thalidomide, anyway, had been tested in rats and rabbits.

In hindsight, researchers misread the clues from the beginning. The birthrate in rats was a little low. But rats have a peculiar ability to protect against defective offspring; the faulty fetus breaks down within the womb, its tissue absorbed and metabolized away. A low birthrate might hint at that—that faulty fetuses were being reabsorbed—but that conclusion could be only be a guess. The rabbits, well, their hind legs were a little warped looking. But just a little. European regulators decided that the risks appeared minor. The U.S. Food and Drug administration was more cautious; scientists there were troubled by some strange inconsistencies. Thalidomide made people sleepy but not rats. Could people somehow metabolize thalidomide differently? While the FDA debated approval—and was loudly criticized by American women wanting the drug—tragedy was waiting. Within two years, some 4,000 badly deformed children were born to European mothers. They were called thalidomide babies. The word "phocomelia" suddenly entered the scientific vocabulary. It meant seal limbs.

That's what thalidomide babies had, stubby little flippers instead of arms and hands, legs and feet. And a few other things: missing ears, eyes fused together, and what medical researchers politely call "an absence of the normal openings of the gastrointestinal tract." In Europe some parents were so appalled, they committed suicide. Others deliberately poisoned their damaged children.

In the midst of the uproar, primate researchers started to wonder. Was the primate metabolism different? Could that be why the drug's impact was so vague in rats and body-destroying in humans? They gave thalidomide to pregnant baboons. There they were again, the flippers and the missing ears. Even today, scientists are not sure how thalidomide manages to twist an embryo so badly. The lesson, though, from young baboons and humans was that there was something in the monkey-human metabolism that turned thalidomide vicious. In rats and rabbits, the drug was a little dubious; in primates, it was unquestionably destructive. Take the polio success, take the thalidomide disaster—the conclusion was that we could not do without primates in biomedical research.

Since the middle of the century, since the rise of jumbo jets and high-powered tranquilizers, monkeys have been plugged into an astonishing array of experiments. In the 1950s, primate researchers developed chlorpromazine, one of the most powerful drugs used to treat mental illnesses such as schizophrenia. In the 1960s, monkeys were used to develop a vaccine against rubella and in the surgical transplanting of corneas to restore vision. In the 1970s and 1980s, primate research helped track down tumor viruses, to improve chemotherapy. The now widely used vaccine for hepatitis B was developed largely in chimpanzees. Many researchers believe that, without primates, a vaccine against AIDS will never be developed.

It was in macaques that researchers learned to do organ transplants. Potent anti-rejection drugs, such as Cyclosporine, were first used in nonhuman primates. The design of the massive heart-lung transplant was developed in rhesus macaques. The connection is so close that surgeons have, on occasion, taken the heart or liver from a baboon and transplanted it into an ailing human. In the mid-1980s, for instance, California surgeon Leonard Bailey put the heart of a baboon into an infant girl, whose own heart was failing. In the early 1990s, Pittsburgh surgeon Thomas Starzl twice transplanted baboon livers into humans, men whose own livers were crippled. In all cases, the humans did not survive more than a couple months with the baboon organs. But the surgeons have announced they will try again.

That's not to say that monkeys are the first line of animal research. They're not—they're too big, too expensive, too wild, too rare to put into all lab tests. About 40,000 monkeys are used annually in American research today, but that stacks up against some 15 million rats and mice. After rats and mice come about 375,000 guinea pigs, 300,000 hamsters, and more than 100,000 dogs. The number of cats used in lab experiments is about the same as monkeys. Control of diphtheria came from guinea pigs and horses. Open-heart surgery was developed in dogs. The critical diabetes work was also done in dogs. From sheep came control of anthrax, from cows the eradication of smallpox. Of course, the use of large animals, like cows, is rare in medical research. The emphasis has been on small domestic animals. They are easier to work with and less costly to handle, requiring fewer precautions than wild-caught monkeys and apes.

The number of primates used in laboratories is down, way down from the 1960s. The investment in primate research, though, is solid. By 1960, the National Institutes of Health decided the animals were so important that it needed special institutions to study them. There are seven NIH-funded, regional primate research centers today, each operated through a university. The centers are run by the University of California in Davis; the University of Washington in Seattle; the Oregon Health Sciences University in Beaverton; the University of Wisconsin in Madison; Tulane University in New Orleans, Louisiana; Emory University in Atlanta, Georgia; and Harvard University in Boston, Massachusetts. The centers receive more than $40 million a year directly through the government's primate research program, in addition to grants from other agencies and private business. There seems to be almost no medical condition that the primate centers haven't explored with monkeys. Research has included investigations of AIDS, cancer, heart disease, development of artificial arteries, aging, spinal cord injury, leprosy, malaria, Parkinson's disease, obesity, nutrition, infertility, in vitro fertilization, birth defects, and baldness.

That is only a partial list, of course. Still, the federal centers represent only a part of the primate research industry in this country. Major independent facilities include LEMSIP and Southwest Foundation for Biomedical Research in Texas, which houses more than

3,500 baboons. A research center at Duke University in North Carolina studies endangered and primitive primates, such as lemurs, tarsiers, and lorises. There are also numerous small university colonies and uncharted lists of private companies, particularly pharmaceutical firms, that continue to use primates for research. In California, for instance, the largest user of monkeys, after the Davis-based federal primate center, is Syntex, a drug company with a major research complex near San Francisco.

Since the early 1970s, the number of full-length publications based on primate studies has doubled, from 2,500 a year to 5,000 a year. Monkeys—due largely to cost and conservation—may be harder to get, but primate researchers are growing more fascinated with them, not less. They are spinning out more information from them, not less.

The development of vision in monkeys and humans is so close that scientists can study sight in monkeys by putting extended-wear contact lenses on their eyes. Their eyes water. So do humans'. The same tiny cilia cells move the tears out of the eyes of monkeys and humans in the same cleansing reaction. The vision of baby macaques and infant humans is almost identical at birth. The monkey's eyesight sharpens about four times as quickly, but that only works to the advantage of animal researchers. Scientists at the Yerkes Regional Primate Research Center, the Emory-run facility in Atlanta, have developed ways to treat human babies born blinded by cataracts, using research gleaned from monkeys.

No one says the brains of humans and monkeys are identical. No one tries to say such a thing, even about humans and chimpanzees, whose DNA matches ours by 98.5 percent, a difference that is obviously next to nothing and next to everything. But still the closeness is there, spelled out in the very vulnerabilities of the brain.

As in humans, epilepsy occurs naturally in baboons, an electrical sparking in the temporal lobe of the brain. Monkeys have turned out to be the best model for Parkinson's disease, sharing a similar collapse of nerve cells in the brain stem. Scientists tried inducing Parkinson's in rats, with drugs, but the rodents shrugged the chemical off. Monkeys, though, developed the shivering loss of control that plagues human sufferers of Parkinson's disease. Also, unlike rodents, both people and monkeys lose brain weight between early and late adulthood. And they lose it in the same places, in the forebrain, the brain stem, the cerebellum. They lose it at the same rate, memory and motor control slowly dying away.

Rumbaugh's computer-game work was focused on the monkeys themselves, how to entertain them in space. That he showed they share analytical abilities with humans was a side effect. There's a ripple effect to that kind of work, though. During the past few years, scientists in California have been tracking the analytical prowess of rhesus macaques for another reason. It emphasizes what a good model they are for the human brain.

An interest in learning and memory attracted Peter Rapp to monkey work. Rapp is a research scientist at the Salk Institute for Biological Studies in La Jolla, near San Diego. He was interested in the aging human brain but he wondered if monkeys, like humans, started out able to conduct sharp analysis and lost that in the fading of the brain's abilities. Like Rumbaugh, he went for the high risk. Could the monkeys infer? Would they be able to sort out a logical sequence of relationships: A is greater than B, which is greater than C, which is greater than D?

Rapp added a twist beyond what Rumbaugh had done, though. In Rapp's tests, the higher-valued symbol always elicited a food reward. That meant that, in the A/B relationship, the monkey had to chose B to get a treat. But B produced no treat in the B/C relationship. In this case, the choice had to be C. So, it wasn't just a matter of recognizing the letter. The monkeys had to also understand its context. They did. Rapp's studies have gone from A to G, and he is looking toward stretching them further.

Rapp wants to take advantage of the fact that the primate centers are now more than 25 years old; many have monkeys approaching that age as well. The aging macaques' bodies are like those of old people—skin softening and wrinkling, the golden-brown hair graying, lost teeth, fumbled coordination. What Rapp wants to know is, do they lose their memories in the same way as people do? He wonders if that ability to grasp the relationship of A, B, and C will fade away.

Like humans, aging macaques seem to first lose their memory for recent events. Then the past starts to slip away from them. Like humans again, they become slower to react, more set in their ways. And yet, Rapp points out, the losses are not evenly spread through the population. Some macaques stay whistle-sharp through their lives, as do some people. Others begin to falter earlier. That variability interests him. What is it that makes a particular brain more vulnerable? If he could sort that out—the toughness of one brain, the fragility of another—he might find a way to shelter those most vulnerable, and block the loss of memory.

"It's not that I see my work as clinical or practical," he says. "It's not. But I do think that if you understand the brain, if you understand memory, at some level you understand who we are. If you're one of those unfortunate people who goes through a cognitive decline, it's not just that you can't remember to take a pill, or that you forget a telephone number. It's changing what used to be you. Memory is at the seat of what makes us individuals and makes us what we are."

He doesn't doubt that's true for monkeys as well.

"It's hard for me to think of something, other than verbal tasks, that one can't find at least suggestive evidence that nonhuman primates do the same thing. It really becomes an ethical dilemma. On the one hand, from a research perspective, it's ideal. On the other hand, it starts to blur, a little bit, the boundaries between animal and human research."

It brings Rapp very close to where Rumbaugh is today. Could you ask rhesus macaques to talk? And what would you do if they did?

RUMBAUGH BEGAN WORKING with monkeys in part because of the Suez Canal crisis of 1957. He had finished his Ph.D. at the University of Colorado and just moved to San Diego State College. He'd barely arrived in California when he was contacted for military service and stationed at the Naval Medical Research Institute in Bethesda. He acknowledges, wryly, that his introduction to primate research was a far cry from his current elegant laboratory.

In the 1950s, when no one was bothering to really argue about Descartes, no one worried too much about clean housing for monkeys. What Rumbaugh remembers most about the Navy laboratory is the cockroaches. Not that the Navy didn't try to get rid of them; technicians sprayed insecticide constantly. Rumbaugh became convinced, though, that chemicals were somehow encouraging the growth of mega-roaches, scuttling about like fat black cigars on legs. "We had marmosets and squirrel monkeys there, and these huge roaches running all over the place. It was dangerous, if you stepped on one, you'd fall and crack your butt. But the marmosets were fascinated. They'd watch them running along the cage bars; then grab them and eat their heads off."

The Navy had some old V-2 rockets, left over from World War II. Officials wanted to salvage them for space flight experiments; it was a time of rushing to catch up with the Soviet space program, by any means available. The rockets weren't that big. But the scientists decided to use them anyway. They would test the force of gravity by putting squirrel monkeys in the nose cones, firing the missiles, and seeing how the animals survived the ride.

Rumbaugh doesn't recall the experiment as being too much slicker than the laboratory itself. When the first rocket fired, they lost the monkey. Literally. They never could find the nose cone. The other monkeys survived the ride though. They also survived being spun on a centrifuge to simulate gravity loading. To keep the monkeys on the centrifuge, Rumbaugh's group mixed up a sticky, plasticky material to hold them still. They buttered the animals to keep them from being permanently stuck; the monkeys breezed through the ride and loved licking off the butter. The Navy was also impressed. It leapt to the conclusion the sticky stuff was protecting the monkeys—and might protect people—and classified it. Later, researchers realized the tiny squirrel monkeys were too light to be really battered by gravity. The one-pound animals were a lousy model for a 200-pound man. "In retrospect, it was a little embarrassing," Rumbaugh says, with wry self-mockery.

It was a reminder that has stuck with him, a reminder that monkeys are never a perfect substitute for human beings in research. They are not the same shape, weight, size. Still, Rumbaugh was hooked. In his mind, it wasn't important for them to be human mirrors. And it wasn't the outer package that interested him, anyway. It was the brain within. He envisioned a continuum of capabilities and intelligence. If you could understand the mental abilities of apes and monkeys, surely that would provide some insight into the evolving intelligence of humankind. He returned to San Diego, then moved to the Yerkes Regional Primate Research Center in Atlanta, and later to Georgia State University. With each move, he became more curious. How smart were they really, our fellow primates? How far, if you were willing, could you push them?

While doing their computer-game studies, Rumbaugh and Washburn found that rhesus macaques reminded them—sometimes a little too much—of video-game-playing high-school students. The monkeys did better on the games that interested them; they blew off the boring ones. And they did consistently better on the tasks they chose rather than the ones a trainer picked for them. "Don't we all?," says Rumbaugh. "But no one knew, before, that it was important to animals."

The monkeys have the quick response of teenagers, too. They learned to adjust the speed of their shots to match the speed the computer assigns a target. They abort bad shots, once they realize they will miss the target. Abel and Baker are also intensely competitive. Side by side, they fire more shots per minute, recall more bad shots, and hit the target more quickly than if each monkey is working alone. If they are separated, not only does their interest in the games fall off, but so does their accuracy. If they can't show off for their friends, it isn't as much fun.

So Rumbaugh watches as the rhesus macaques carefully choose the video games they like best. He watches them ace the tasks they like. He finds the conclusion inescapable: "The only reasonable conclusion is that it's a rational process."

There are rumbles—and Rumbaugh has heard them—that we don't really want to know these things. Keep it simple, keep it easy. The more science reveals about the intelligence of other species, the more difficult the questions about using them. Once you consider them to be rational creatures, analyzing their relationship with the world and with other beings around them, the picture changes. It is less easy to regard an animal as a tool, another resource on the shelf, a furry test tube, as Roger Fouts complains. "People may think our work introduces complications into research," says Duane Rumbaugh. "I have no apologies for that. I know that scientists want to conduct their research responsibly. They will do so, given a proper understanding of their primate subjects."

But while Fouts crusades passionately for the rights of animals, Rumbaugh does not. He is not a preacher. He believes the research itself carries more weight than words, that the documented abilities of Abel and Baker are as effective as any speech in reminding researchers that the territory is changing, that this is no Medieval France, no longer the world of René Descartes.

THREE

The Black Box

THE MOST DIRECT way to explore the head of a monkey is with a power drill and a scalpel. Open the skull. Burrow into the brain.

If you're interested in the machinery of the brain, you go straight in like that. A scientist like Duane Rumbaugh, chasing intelligent behavior, has no need to pry open the skull. He wants to watch the monkey in action.

It's different, though, for a researcher interested not in the monkey as a whole animal, but in one crucial part—the brain itself. The body protects the brain well, buffering it with layers of armor, concealing it. A neuroscientist, an investigator seeking the inner workings of the brain, must go inside it to understand it. No technology yet available, no imager, no powerful combination of magnets and computers, can yet see the live brain in all its intricate and ever-changing detail. If you really want to know, then you make a path through blood and bone. Otherwise, if you are too squeamish, you walk away from the secrets within the skull.

People complain that there is no unexplored, unmapped territory left in the world. A neuroscientist knows that's not true, that the most mysterious territory left on earth lies within the skull. In the operating room, you begin to understand how mysterious. It takes tedious hours to get to the brain; a mind-numbing repetition of cut and snip. Yet still, every time, there's this moment when the barriers are cut away—the skin, the muscle, the skull, the thick, gray membrane that lies beneath the bone. The folds of the brain show through. They are gleaming, wet, shell-pink, coiled into fine loops. In that moment, a monkey brain shines as brightly, offers as much challenge as a human brain.

Neuroscientists have been chasing that challenge forever. Or so it seems sometimes to Stuart Zola-Morgan, a man who has given his career to coaxing secrets out of the unknown territory. Zola-Morgan is a research scientist in psychiatry, a neurosurgeon at the University of California, San Diego. He also holds an appointment at the Veterans Affairs Medical Center, at the UC-San Diego campus. He chairs the animal research committee of the national Society for Neuroscience, a group dedicated to defending animal research. He's a lover of good jokes, magic tricks, old movies, and the brain. His operating room is a long way from Rumbaugh's Georgia laboratory, in miles and in approach. It's a long way from Roger Fouts's laboratory too. This is not research that puts the animal first at all costs. The brain is the holy grail here, and the monkeys are among the tools needed—vital and valuable—on the quest.

When Zola-Morgan explains the magic of the brain to his students, he calls it "the black box." His hands shape the box in the air—not so big, grapefruit-like. He makes a simple curve, revealing nothing of the webbed complexity inside. "I tell graduates that it's as if whoever runs the show says, 'Look, here's this black box. There's lots of complications in how it works. And if you spend endless days, and nights and years, I may reveal to you just a little about how it works. I may let you have a tickle.' So then I look out at the students and I say, 'Do you still want to do this?' " He starts to grin. "It weeds them out." He thinks about it a minute himself, graying brown hair crammed under blue surgical cap, brown eyes thoughtful behind steel-rimmed glasses, gloves peeled off for a break, hands freshly scrubbed—and steady. He's been chasing the mystery for decades and his interest hasn't flagged. "Even the tickle is so amazing."

It is because monkeys are so like us—their intelligence so tantalizingly similar to our own—that they have become such a good model of the human brain. Zola-Morgan uses monkeys because he can't answer his questions without them. He uses them because they give good answers, because in trying to decipher the mysteries of their brains, he gets closer to understanding ours. He believes without question that the pursuit is worthwhile, that the use of monkeys in his research pays off. It does more than feed a fascination; it promises knowledge that can be used to help people with brain damage, afflictions such as Alzheimer's disease, and crippling strokes.

If you follow his work through even a single surgery, the tradeoff as he sees it becomes clear: There is a price to be paid for knowledge. In this case, the price is in the lives of another species.

Surgery days begin early for Stuart Zola-Morgan. On a cool January morning, the mist still wraps silently through the dripping palms, fig and orange trees, callas and bleeding heart ferns that tangle around his home in south San Diego. Inside, the scene is anything but quiet. Rhett, the Yorkshire terrier, and Lizzie, a mutt, bounce off the door, begging to go out. Zola-Morgan's wife, Susan, a veterinarian at UC-San Diego, brews coffee, easing into the day. Coco, the cat, snoozes on the floor.

Zola-Morgan is too edgy to sit still. He stands by the window, looking out into the garden, but the wet leaves just blur away. In his mind, he sees the operating room. He knows that at the campus they're already prepping the monkey for surgery. Sitting for breakfast seems intolerable. He grabs a cup of coffee, snatches a package of Pepperidge Farm Mint Milano cookies, climbs into his red Toyota, and sends the car plunging through the mist, down the steep hill of his driveway.

It's like being a fish on the hook. On mornings like this, Zola-Morgan's work reels him in, the tug of an invisible line between his operating room and his bed. His current series of experiments focus on some of the brain's smallest, most elusive structures. They seem to demand the inner buzz, maybe even the jolt of coffee and cookies. He can guarantee himself a minimum of eight hours in the operating room, below ground, in the basement of the UC-San Diego Medical School. Gowned in blue scrubs, ringed by white-tiled walls, shiny with disinfectant, his gloved hands smudged with blood, he will slowly, carefully, cut his way down to the monkey's brain.

Getting there it is always a lengthy procedure, at once tedious and grisly. On these mornings, Zola-Morgan arrives at his campus operating room at 8:30. He must wash up, don his scrubs, hat, booties, two pairs of sterile gloves, before he even approaches the waiting monkey.

He uses crab-eating macaques, a close relation of the rhesus. The monkeys are less scrappy than rhesus macaques, less hardy. This day's monkey had come, months before, from the Philippines. It would be already motionless on the steel operating table, eyes closed, deep in an anesthetic-induced sleep when the doctor gets there. As youngsters, crab-eating macaques sport tufts of hair that poke from the tops of their heads; they look like monkeys with Mohawk haircuts. For surgery, the macaque's head is shaved clean.

Zola-Morgan's first job is to brace the head. There is no wiggle room in brain surgery. He opens the monkey's mouth and inserts a metal T-bar, the top of the tee catching behind the sharp canine teeth. Two metal prongs, blunt-ended, grip the area just outside the edge of the animal's eye sockets. With teeth and eyes anchored, the monkey's head is held rigidly still, awaiting the first cut. Holding a scalpel, Zola-Morgan traces a T-cut onto the top of the head, a line behind the brow-bone, a perpendicular cut down the back of the head. Where the knife touches, blood glimmers bright red, showing through the skin like the pulp of a fruit. He rolls the severed skin away from the muscle.

He has learned to risk a little extra blood. Ten years of operating on monkeys prejudiced him against cauterizing the tissue to control bleeding. The brief burn of electric current rapidly clots blood, but it cooks cells as well, slowing the healing in the muscle. He prefers anesthesia that lowers blood pressure, slowing the flow, indirectly reducing bleeding. He learned the trick from neurosurgeons who operate on people. He asked them for ways to make the surgery less traumatic.

But the dense tissue over the monkey's skull—the powerful muscle that give its jaws such strength—is about ten times thicker than the comparable muscle in a human head. It is crammed with blood vessels. The cut muscle inevitably runs red. As the surgery progresses, gauze pads, soaked with blood and saline, pile up around the scientist's feet.

And the tough muscle holds hard and tight to the skull, resisting his efforts to pull it away. It is a good hour until he can see the left cheekbone, the pinkish-white curve of the zygomatic arch. Still, he has to free it further, hitching a square of gauze behind the bone, dragging it up and down until the muscle is polished off and the arch stands clear. Then, using a drill that fills the room with the whine of metal on bone, he breaks out the arch. He cuts on, opening a path through the muscle to the skull. Then more drilling, making a walnut-sized hole in the skull itself. Then, carefully, he chips away the bony skull to expose the tissue that covers the brain, the thin, gray rubbery-looking sheet of the dura. And then the whole process must be repeated on the other side of the head.

The surgeries have been known to tick away for so long that Zola-Morgan's graduate students, working with him and his colleagues in the operating room, despair of ever getting a meal. Sometimes they sneak out, holding their sterile gloved hands in the air, opening their mouths like young birds while their friends stuff in chocolate or pretzels.

And all of that is mere preparation, what must come before starting on the main task of the surgery—playing with the brain's internal switches.

Now, the approach becomes not unlike that of a new homeowner exploring his circuit-breaker box, trying to figure out what controls what. Flip this switch and the kitchen lights cut out. Flip that one and there goes the living room, the garage, the bedrooms. What Zola-Morgan tries to do is to turn off parts of the monkey's brain and then see what happens to its memory.

The research team knows from experience that you can train macaques to pick up objects in a certain order: first the green block, then the red. It's a simple task of memory. The question is, will a monkey remember that order if Zola-Morgan has removed part of his brain? What part? Will the monkey remember that green came before red if you flip off this switch? That one?

What if you probe into the brain, down into the temporal lobe, just in front of the ears, to the place where Alzheimer's disease begins its ruinous journey through memory? What if you probe right to the hippocampus, a gray ridge of cells curled deep into the lobe? Named for its distinctive, seahorse shape—"hippos" is Greek for horse, "kampos" for sea monster—the coil of cells floating in a sea of neurons has been thought to be the gatekeeper of memory. It has been suspected of being the thing that chooses, that says "Save this thought. Let this one go."

So if you carve it out, what then? What if you chiseled away the brain tissue on either side of it, leaving gaps, dark spaces in the brain? What would be lost? What would stand?

Driving from his home, crammed amid the morning rush of Southern Californians on Interstate 5, Zola-Morgan anticipates the surgery, hoping he will get to ask those questions. He knows, as it has happened before, that the operation could fail on the table. It takes a good four hours just to open up the brain, bracing the monkey on its stomach as scalpel and drill bite deeper and deeper. Even unconscious, the monkeys' bodies are vulnerable to the shock of invasion, so much so that sometimes just turning the animal over will cause its heart to stutter to a stop.

The researchers try to prevent it, using only the healthiest, toughest animals, running blood tests before the surgery, checking against signs of illness. The monkeys can fool them. The animals are born to hide illness. A crab-eating macaque, its instincts tuned to the rainforests of Asia, has good reason to suffer in silence. In the wild, an obviously ill animal is a target. A cry of pain is a siren call for predators. The crab-eaters grow up learning to hide their vulnerabilities, including those that make them a bad risk for surgery. Ketamine, the popular drug used to immobilize animals before surgery, will overwhelm a diseased liver. Zola-Morgan has lost animals for just that reason. The monkey never showed a symptom until the drug was administered.

Out of the 70 monkeys used in his experiments during the past decade, a half-dozen died in surgery. He has become religious about doing blood tests before surgery, checking the function of the liver. He guards, as best he can, against the loss of an animal.

He knows too well that even if the operation succeeds, the brain will not suddenly open up, unfolding like a flower in spring. The brain is unforceable; it will be understood sliver by sliver, if at all. He has resigned himself to that. He hopes to live a long time. He doubts it will be long enough to answer his questions. The unyielding intricacy appeals, in a curious way, to the clockmaker traits in him, the part that likes to go hands-on with gears, cogs, and wheels. "My mother says I was always doing this when I was small, taking things apart."

He'd started out, at the University of Massachusetts in Amherst, planning to study human brains directly. But the clockmaker in him didn't like the kind of experiments you could do with people. They were too unpredictable, too blurry with emotional response, too messy. "I started to become a little disenchanted," Zola-Morgan says. "There was a lack of rigor. I don't mean that in a negative way, I just wasn't as comfortable with it as others were." He was pondering his future when Harvard University ran a want ad in the local paper. "It said something like, 'Interested in monkey business? Call this number if you'd like to study primates.' " Curious and restless, he called. The Harvard group was studying high-fat diets, taking baby monkeys by Caesarean Section and hand-rearing them on a formula diet. The little monkeys were unexpectedly showing signs of stress, rocking themselves, sucking their toes constantly. "It took someone with basic Psychology 101 to know they needed interaction with other monkeys, they were too isolated," Zola-Morgan said. But from that start, he became interested in monkey behavior, and then interested in the physical processes that drove the behavior.

"I just fell in love with the brain. This organ that was so complicated and so mysterious and so responsible for our behavior. And what really interested me was memory. That seemed to be at the core of who we are. The funny thing is that, back then, we thought memory, at least, was simple. It was very straightforward. And I've spent my career making it more complicated, finding that there's more than one memory system, that they operate by different processes, that they're housed in different structures. The harder you look, the more remarkable a mechanism the brain is."

Neuroscientists have dubbed the 1990s the Decade of the Brain, the era when new technology and new insight will finally allow humans to understand the machinery inside their heads. Sometimes, they are sure that will happen. Sometimes, they think even astronomers have an easier job. Astronomers analyze stars just a few hundred billion miles away; they merely pick apart the composition of some distant blur of light. Researchers such as Zola-Morgan are trying in a sense to turn themselves inside out, to use their brains to understand their brains. It seems, sometimes, like the last test of human intellect—to understand itself.

That sense of testing—and being tested—runs through the culture of the field. You can hear it in Zola-Morgan's words, in the thoughts of his colleagues. David Amaral, a neuroanatomist who has worked closely with Zola-Morgan, calls the brain "not just the most complicated organ in the body, but the most complicated product of evolution." Would that it were simpler, he says; scientists have done their best to make it so.

Amaral moved to the State University of New York at Stony Brook in late 1993, after collaborating with Zola-Morgan in San Diego. In California, he worked at the nearby Salk Institute for Biological Studies, down the road from Zola-Morgan's office. When Amaral became interested in the internal workings of the mind, back in the 1970s, researchers were cheerfully subdividing the brain into centers: pleasure center, pain center, and so on. Amaral doesn't deny the brain is specialized, but he dislikes any attempt to oversimplify it. The whole idea of dividing the brain up into nice bright boxes annoyed him so much that it drove him out of psychology, his first career choice.

He wanted to understand the structure of the brain, brick by brick. He became a neuroanatomist, a brain mapper. Other neuroscientists sometimes poke fun at anatomists—the nerds of the brain, the guys who get fixated on how slot A tucks into section B. Amaral, a blue-eyed, soft-voiced New Englander, liked looking at the brain, seeing it as something solid, something recognizable. He liked the scenery, looking for a particular nerve as it wound through tissue, finding it. It was real. Like Zola-Morgan, he enjoyed sharing that sense of discovery, working with students. Peter Rapp, who is so painstakingly testing the memory of crab-eating macaques, is one of Amaral's students.

Like Zola-Morgan, too, Amaral finds a certain appeal in the cogs-and-wheels aspects of the brain. Since the turn of the century, neuroscientists have moved from believing that vision was processed by three parts of the brain to suspecting the right number is between thirty and forty parts. Amaral hopes to help do the same for memory, to find the bits and pieces. It is a daunting task. At least vision researchers can start with the eye and work their way inward. The brain processes many different kinds of memories, from many different sources. Even the question of where to begin looking is intimidating.

"We no longer think that we're going to put it all together in five easy pieces," Amaral says. "That might have been the conceit 20 years ago. But these days, I think we're a lot more respectful."

Computers and artificial intelligence have only reinforced that respect. Consider the ordinary task of picking up an egg: You must see the egg, remember what it is, send the proper signals to your hand to pick it up, and not just to grab it, but to hold it with the softest touch, so that the thin white shell doesn't crunch in your hand and send a stream of slimy white and yolk down your fingers. Engineers trying to design robots to recognize and pick up objects have destroyed a lot of eggs. They have found that it takes the machines hours to do what the brain does in a split second.

Further, the brain processes a vast amount of information about the egg—white, cool to touch, slightly rough on the surface, shaped in an irregular oval. Simultaneously, while holding an egg, memories may come rushing in: of hunting dyed eggs at Easter, the bright colors flashing between tree roots and under bushes; memories of deviling eggs for a summer picnic; of searching in the hen house for eggs still warm from feathered bodies; of breaking an egg onto the heat of a sidewalk in midsummer; of walking away from the uncooked mess.

It's the chain of memory that obsesses Zola-Morgan and his colleagues, or rather, how the brain holds and retrieves separate understandings. Where does it store its memory for pictures of eggs, visual memory? Where is its recollection of how they feel against the fingertips, touch memory? How about its memory for sneaking through the dark on a Halloween night, carrying a clutch of eggs to break on windshields, autobiographical memory? And there's short-term memory: following a cake recipe, remembering that you've already stirred the egg into the batter. There's long-term memory for facts: whipping eggs in a copper bowl because you read, several weeks ago, that the whites fluff better against that metal. There's conscious memory: the ability to say yes, I remember, I bought eggs yesterday. Unconscious memory: the unthinking, automatic feather-grip of fingers around the white thinness of an egg shell. Incredibly, each of them winds like thread from a different part of the brain, weaving together into a fabric or spinning out as a separate fiber, according to need.

How does the brain pull all those recollections out of their proper place? How does it so precisely bring them together? Never mind the whole brain, can anyone understand just memory? Zola-Morgan wonders. So does Amaral. So does the third member of the team, Larry Squire, a research scientist, who like Zola-Morgan holds joint positions at UC-San Diego and with the VA hospital. They aren't even trying to hunt in every corner of the brain. The San Diego group has focused on conscious memory, searching for it in the temporal lobe. The scientists have begun to find the chain, the linkage between daily events and long-term memory, the brain's storage system. It has taken them years. Sometimes Zola-Morgan believes that patience and stubbornness may be the most critical elements of success in science. He's not sure which to rank higher.

Think of the brain, for a minute, as an enormous kitchen pantry occupied by a hyperactive octopus. Shelves stretch up into the shadows, crammed with eggs, tins, boxes, recollections of things past. Tentacles whirl to retrieve these things, several at a time, at blurring speed. Then think of the scientist as a kind of kitchen spy, on his knees, peering through the keyhole, and from that narrow field of view, trying to understand what's happening inside, even when the octopus' arms are reaching to shelves he can't possibly see. And then add the complication that sometimes the octopus is flawlessly efficient and sometimes it isn't. Sometimes, after all, people pick up the egg and do a blank: Where am I? What am I doing with this? And sometimes worse, as memory crumbles with age or illness, they hold the egg, staring at it, wondering, is this something I've done before?

No wonder that, in some ways, opening up a monkey's brain has such appeal. It's like prying open the pantry door. But, like good spies, the scientists cannot just go barging in, smashing up what they want to study. The patterns of damage that Zola-Morgan and his colleagues etch are not simple, random destruction. They are mimics, in a sense, of injuries that have occurred in human brains, of injuries that have devastated human lives. The hunt is indeed partly the purest quest for knowledge. But there is a more painful edge to the quest. Zola-Morgan's labs and offices may be in a basement wing of the San Diego medical school. They may seem all fluorescent-lit corridors and blank walls, isolated, physically, from the real world above, no hint of the sea winds, sun and trees, students laughing on campus pathways. But they are connected. The studies are driven by a passion not only to understand the destruction of human memory, but to prevent it. These researchers know how unbearable a loss it can be.

Larry Squire has spent his career trying to understand, and perhaps one day help repair, human amnesia. From those years, two men stand out in his mind—one still alive, one now dead.

The first is H.M., known publicly only by those initials. It's a scientific alias, a buffer to protect the privacy of a research subject. In the 1950s, H.M. was an electrician's assistant with a vicious case of epilepsy. He was trapped in a cycle of daily, bone-cracking seizures. There were no good drugs for epilepsy then. Desperate for control—even a day of controlling his own body—H.M. sought medical help. He picked an unfortunate time. Those were what Squire calls the psychosurgery days. Doctors were experimenting with lobotomies to cure schizophrenia. If the brain wasn't working right, they would go in and rearrange it. Epilepsy was known to center in the temporal lobe; the electrical crackle of seizures was thought to somehow begin sparking there. So H.M.'s doctor opened up his skull and slid a scalpel into the trouble zone.

The knife cut directly across the seahorse shape of the hippocampus, leaving its head, taking out midsection and tail. It destroyed a tightly wound cluster of cells nearby, called the amygdala, from the Greek word for almond, which it resembled in shape. And it meandered through a good three inches of tissue surrounding the hippocampus and amygdala, at that point basically unknown territory.

As an epilepsy cure, it was extremely effective. The seizures, except for the rare tremor, vanished entirely. The price? H.M., now in his late sixties and a longtime resident of a nursing home, has described his life since then as the continual sensation of awakening from a dream. He remembers the 1940s, the shock of Pearl Harbor, the deepening war, the loss of President Roosevelt. But by the following decade, the past blurs away, even the past of half an hour ago. People who, as they grow old, seem to re-live their youthful years are described poetically as "suspended in the past." Larry Squire describes H.M. as being suspended in the present.

From accounts by scientists and other observers, it's clear H.M. can read and understand an entire newspaper, but 20 minutes later he recalls nothing from it. One of his favorite tasks is crossword puzzles; as you work your way through them, you write down your answers, leaving a permanent record of what has come before. If anyone had doubted that the temporal lobe was critical to memory, they had only to come to the nursing home and spend an afternoon with an aging electrician's assistant.

"He's the index case," Squires says. "When people first heard about what had happened, it launched an incredible battle in science. It implied a tremendous separation between intellectual capabilities and memory. The idea was that if you were smart, paid attention, you remembered well. And obviously with H.M. that didn't hold up."

H.M. can still, after all, analyze a newspaper. He can be taught new skills. Scientists have taught him to "mirror-read," comprehend mirror-reversed words. He can do it swiftly now. But he cannot remember being taught the skill. A researcher may spend a morning asking H.M. questions. If he returns after lunch, H.M. stares blankly, recalling neither questions nor questioner. Events slip through his mind as relentlessly as shadows driven by the sun's chase across the sky. But there are also abilities he has been spared. He retains extreme short-term memory, the kind that allows people to briefly remember a phone number. Skill memories are intact. He remembers in detail the past before the surgery. And he has remained, as he apparently was before the operation, a sweet and patient man.

The second man on Squire's list went by the initials R.B. In 1978, the 52-year-old man underwent open-heart surgery. His heart faltered in the recovery room, choking off the flow of blood to his brain, triggering a rapid, massive stroke. He survived another five years, struggling to piece back together a shredded memory. He had not lost as much as H.M. He had a better hold on the recent past. He could remember people better. He was unable though to learn a new task, to hold onto a new fact. He knew enough to realize what he had lost, and he wanted something good to come of it. He promised his brain to researchers, agreeing that, when he died, they could open his skull and hunt down the precise damage in his brain. When they did—after receiving a predawn call from his wife in the spring of 1983—they found a surprisingly small scar, barely a white streak across the hippocampus.

From these cases, brain researchers gained powerful hints about what was happening in the temporal lobe. H.M.'s clue was that long-term memory was somehow centered there, in the region of the hippocampus. "Everyone thought the hippocampus was just magical," Squire says. R.B., whose lesion was so neatly centered in just that structure, showed that there had to be something else. It couldn't be the hippocampus alone; if it was, R.B. would have been as lost in time as H.M. Obviously, there was something in the brain tissue around the hippocampus, that extra knife cut that H.M. had suffered, that was also vital to holding memories in place.

The first suspect was the obvious one, the amygdala, so close to the hippocampus, so clearly a defined structure. And then, the San Diego group found itself focused on the unobvious suspects, thinking about the bland, dull-looking surface of the brain just above those structures—the wrinkled stretch of the cortex. If you stare at the cortex hard enough, the region just above the hippocampus and amygdala resolves itself into three separate regions, just a little different in color, the terrain of the cell structure a little more rugged in one place than another. Under a microscope they take on distinctive shapes—the crescent curve of the perirhinal cortex, the slightly boxy shape of the entorhinal cortex, and the rounded outlines of the perahippocampal cortex.

Squire, Zola-Morgan, and Amaral, three believers in complexity in the brain, made a pact to sort out memory in those shadowy corners of the temporal lobe. Each brought a different strength. Squire brought his abiding interest and many years working with human amnesia patients. Zola-Morgan brought his detailed, patient, careful work in the convoluted brains of monkeys. Amaral brought his fascination with the roadways of the brain. If they had a crusade at all, it was to undo the myth that there was wasted space in the brain, that much of it was simply packing material around the good stuff.

Squire, the current president of the national Society for Neuroscience, is sometimes astonished that anyone still believes that the brain ever rests, in any region. He's plagued by a popular mythology, that people only use a fraction of their brains, maybe 10 percent or so. If he asks his graduate students how many have heard that, routinely at least a third of them fling their hands into the air. He suspects that it is almost like a faith, out of the human tendency to hope—to hope that we can do better, that there are untapped regions of the brain that, once we find them, we can somehow harness and expand ourselves into a global community of Albert Einsteins.

"Not that we can't do better anyway," he adds, smiling. "But there is no unused part of the brain. The problem is, we just don't know enough about the brain. It's not just that we haven't figured out all the different functions of memory. It's that basically we know so little about the brain and what it does."

As they deepen their realization that memory may be one of the most complicated systems within one of the most complicated systems on earth, scientists find themselves in a certain spot. They need to look at a brain that can coil and curl and wrap its tentacles around memory. A pantry without an octopus is of little use to them. So they've turned to monkeys. "All brains are complicated," Zola Morgan says. "We don't have any corner on the market in that respect. It's just that the level of complexity is closer in monkeys and humans." The field relies on monkeys; there are few human volunteers for brain-injury experiments. Zola-Morgan admits that he couldn't do the work if monkeys weren't so smart. "I hesitate to talk about how smart monkeys are because it tends to feed the animal rights movement," says Zola-Morgan. "But as a scientist you have to honor what you see. I wouldn't say monkeys have simple brains. Not simple at all."

Of course, even the simplest brains remember. Fat, slimy sea slugs have memory, on a very small scale. The sea slugs have 18,000 neurons, or brain cells. There are an estimated 2 million neurons just in a rat's hippocampus, some 7 million neurons in a human hippocampus. The number of neurons in the whole human brain is still a matter of fierce debate among scientists. A minimum estimate is 10 billion, at tops perhaps a trillion. Still, the 18,000 brain cells of the tube-like sea slug remember. Elegant experiments by Erik Kandal, at Columbia University in New York, found that the first time a scientist gently touches the siphon of a sea slug, the organ it uses to pump seawater, the animal cowers as if attacked, retracting its gill. But after continued gentle touches, it no longer retracts on touch, it "remembers" that the soft tap of a Q-tip, for instance, will not harm it. And if well-trained, if brushed by many benign Q-tips, it will hold onto that lesson for weeks.

It is in vertebrates, however, that structures such as the hippo-campus first appear—in lizards, birds, rats, monkeys, people. In the lines of evolution, the hippocampus is an old structure, evolving be-tween 200 million and 300 million years ago. It remains relatively enormous in brains of lizards and rats, proportionally smaller in hu-mans and primates, where complex analyses, arguments and reason-ing engage the major part of the brain. In all animals who have it, it is clear that the hippocampus is a player in memory.

In rats, researchers can search out the role of memory by drop-ping the animals into a Morris water maze. A metal tank is filled with liquid deep enough to drown a rat and solidly opaque. Water clouded with powdered milk works, as long as the rat can't see through it. In one part of the tank is a small platform, just beneath the surface. A rat is put into the water. He splashes and paddles, searching desperately for a safe resting place until he blunders into the platform. Then the scientist takes him out, gives him a break, and puts him back in. A brain-healthy rat figures the game out within a few dunkings, stroking for the platform in a most business-like way. An animal with hippocampal damage, however, may re-peatedly blunder around, not remembering where the platform lies.

But what does that mean in terms of human memory loss? With monkeys, researchers can basically ask them some of the same ques-tions they ask people: "Do you remember if you saw this (this apple, this tree, this egg) before?" They can't whip out a test form and ask a macaque to fill in the blanks, but they can train it in memory tasks and then test its ability to keep those instructions in place. One way is through a task known as a delayed nonmatching sample. It works like this: You show a monkey a blue ball. Under-neath the ball is a treat—some raisins, some peanuts, or the ultimate for monkeys that need a little extra incentive, M&Ms candies. In the next task, the monkey is shown the blue ball again, along with a rectangular red block. The treat will be under the red block, the object that doesn't match whatever he saw before. You add more and more objects into the formula—pink triangles, green bells, tin-ker toys, stuffed animals, whatever. The rule stays the same, the treat is always under the new object.

Eventually, through repeated training, monkeys learn the rela-tionship. They learn that to be rewarded, they always must pick the new object, the one that doesn't match the object in their memory. So, the tests actually get at some fairly complex thought processes—not just "Do you remember what you saw?" but "Can you analyze its relationship to other objects that you've seen before as well?" Basically, it's conscious memory—the same as, "I remember that I saw that egg before." Squire likes to call it declarative memory, because you can declare what you remember. The test also gets at long-term memory: How long does the brain hold the information?

Scientists can ask people exactly the same questions, and with nearly the same kind of devices, using cue cards, showing red blocks, blue balls, varied symbols. But they've also tested people with amnesia by showing them the blocks and balls from monkey tests. People or monkeys who sustain damage in the region of the hippocampus blunder through the tests. They may remember what they've learned before brain injury. They struggle desperately with tasks taught afterward, especially if delays occur between each choice. Monkeys with temporal lobe lesions can still perform normally if the delay is no more than 8 seconds. But compare a red block with a green ball at intervals of 10 minutes and it's a blank to them. They're like H.M., smiling doubtfully at a scientist who left 2 hours ago and then returned.

You can look physically at the brains of humans and macaques and see the differences. The human brain weighs about 1,500 grams, roughly the size of a grapefruit. The macaque brain is about a tenth of that in weight and size, a plum. Beyond that, the two brains are paced differently, developing at startlingly different speeds. When a macaque is born, its brain is already two-thirds the size it will reach as an adult. A newborn human baby's brain is only one-fourth the size it will achieve. Scientists have tried comparing the size of the brain to the animal's body weight; they call this the encephalization quotient, or EQ. The higher the number, the higher the ratio of brain to body. Humans rate about 7.5 on the scale; apes and monkeys hover at about 2. A rat scores a mere 0.4. To directly compare to a human, monkeys would suddenly need brains three times larger than they actually possess.

There's another interesting comparison, the balance of power within the brain itself. Consider the cerebral cortex, the outer layer of the brain and the layer most recently evolved. In humans, the cerebral cortex makes up 80 percent of the brain. But monkeys are close, at about 68 percent. Neurobiologist Terrace Deacon, of Harvard University, has carefully compared the major regions of the brain in humans and other primates. The prefrontal cortex, for instance, which handles production and understanding of language, is twice as large in humans as in other primates, when adjusted for total brain size. The part of the human brain dedicated to motor skills is only about one-third of that found in the rapid moving, agile monkeys. "As a scientist you try for the best model possible," Squire says. "The monkey model for declarative memory, based in the temporal lobe, is superb." Try to search out another memory, he admits, and monkeys might fail. Unlike declarative memory, tucked in different areas of the brain, throughout the brain, memory for past events is carried in the frontal lobes.

Somewhere in that part of the brain, so much larger in people than monkeys, we store our own autobiographies, our life stories, and our sense, not only of having a past, but of moving into the future. "No one knows how to ask a monkey if he remembers being taught a task," says Squire, "If he remembers the researcher coming into the room, moving the ball. No one knows how to ask monkeys if they have a sense not only of past but of future, of their own mortality. And we have to wonder if they do. These are not animals that bury their dead. But it is clear that these animals do have declarative memory, they do have something like conscious recollection. The data tends toward giving more credit to them than we did before. Not less."

And in the geography of the temporal lobe, the monkey brain is a mirror, in miniature, of the human brain. There is the hippocampus, the amygdala, the subtle, shifted terrain of the entorhinal, perirhinal, and perahippocampal regions of the cortex. The connections are tight enough that doctors who treat human patients have begun looking at results in monkeys.

Zola-Morgan, for instance, artificially recreated R.B.'s injury by mimicking a human stroke in some of his monkeys. It was clear that massive strokes often were based in an abrupt loss of blood and the life-giving oxygen it carries; a rapid starvation of the brain's tissues. So the scientists decided to wrap an inflatable cuff around an anesthetized monkey's neck, tightening it slowly until the blood flow from the carotid artery was flattened to a trickle; the brain was starved of blood. They discovered that if they kept the cuff on for 15 minutes, they created a remarkably specific area of damage, a lesion across the hippocampus.

It took longer to damage other parts of the brain. Somehow, as the study made clear, that part of the brain, the gray cells that make that seahorse-shaped ridge, are terribly vulnerable to loss of blood, more dependent on its flow than any other region of the brain. The rest of the brain gasped, then drew on the new blood when the flow was restored, and got back to business. The hippocampus was massacred. Overall, more than a third of the structure was wiped out. In some parts of the hippocampus, 80 percent of the cells were permanently stricken. The researchers realized with shock that they were actually seeing R.B.'s injury, that it was the vulnerability of the hippocampus that made at least some memory so vulnerable to stroke.

With this knowledge, neuroscientists now hope to jump right into the brain and try some chemical interference. In both rats and monkeys, they've found that if the brain is preloaded with protective drugs before the cuff snaps tight, the damage can be prevented. Perhaps surgeons might one day take that approach in operations such as heart surgeries, where blood supply to the brain can be vulnerable. On the other hand, most people don't sit around anticipating strokes. The attack comes out of nowhere. The work of neuroscientists like Zola-Morgan offers hope here, too.

It turns out that hippocampal cells do not immediately self-destruct. "There's a window of 24 to 48 hours where you could potentially intervene," Zola-Morgan says. "Reduce the level of cell death. Of course, there's still the remaining question of whether preserving the cells preserves function." But the promise is enough. Drug companies are now testing compounds that salvage hippocampal cells. He thinks that within the next decade, drugs will be available to move in rapidly when a stroke occurs, throwing chemical buffers into the brain. Perhaps they will preserve memory.

"It sounds like a fantasy, but it's a realistic one," he says. Yet he wants to make it clear that his own motivation doesn't come from trying to find new treatments. His attention isn't focused on the purely practical at all. "Part of what's driving me is not the fact that I want to develop a cure for brain damage. It's just that I want to be able to understand how this all works. I hope others will be able to use what we find in a practical and clinical way. But the major reason I'm doing this is that I want to know."

He thinks of it too, sometimes, as he spends hours peeling away the protective layers above the monkey's brain—skin, muscle, bone. He thinks of it in the hours that follow, as he and Amaral collaborate on delving into the brain itself. It is Amaral's job to make the actual lesion in the monkey's brain, find the right structure, the fine shadowy outline of a region of the cortex, and take it out. If Amaral is foremost a brain mapper, he is also a surgeon both at ease and extraordinarily careful in moving through the most fragile parts of the brain.

The brain is nothing to go blundering into. Amaral arrives in the

operating room carrying the official brain atlas for crab-eating macaques and a folder stuffed with magnetic resonance imaging scans
of this particular monkey's brain. It sounds almost like a joke, but
no two monkey brains are alike. They share this, too, with human
brains; the brains differ individually, subtly, each one a shading of
tissue apart. The hippocampus may sprawl slightly differently, the
nerves may coil differently, the cortex above may be unexpectedly
deeper, tougher. The whole brain itself may have a slightly altered
balance within the shell of the skull. There have been times that
Zola-Morgan and his students have set aside their brain surgeries to
spend weeks making pictures of monkey brains, implanting tiny
beads as reference points, comparing what lies beneath each fur-
covered skull. They've found that just among crab-eating macaques,
the shape of a brain can shift by as much as 20 percent from monkey
to monkey. Amaral, making his pathway through the brain, does
not want to find himself suddenly slicing into the wrong part.

So maps are the first defense against disaster. The second is a
stereomicroscope. It allows Amaral to focus on tissue in almost a
cell-by-cell journey to his target. The microscope is suspended above
his head, mounted on a high metal pole set in the floor, leaving his
hands clear to search and cut. He gazes into the eyepiece. The view
through the lens is disconcertingly narrow, a slice of pink cells with
no obvious connection to the rest of the brain.

But the third defense, and probably the most critical, Amaral says,
is the three-dimensional diagram he carries in his own head, the
vivid picture of the brain that is always with him. As he peers
through the microscope, as his hands thread the probe down into
the tissue, he is constantly calculating where he is. He thinks of it
as pattern recognition. Or color. The entorhinal cortex is suddenly
paler than the adjacent regions, its nerves plastered heavily with
myelin, the white, dense membrane that provides a buffer. He reminds himself of the amount of pressure it takes to move through
the different layers of tissue, moving slowly. In some ways, this
peephole view of the brain is more relaxing than the global one. He
sees only a few cells at a time; it protects him from the whole picture. When he performs a surgery with the whole brain open before
him, suddenly nerves and tissue shiver beneath his fingertips. The
cuts he makes are no more destructive, but they somehow look
worse. He starts to wonder how he will ever make it to his target
safely, without crashing about like a vandal in a warehouse full of
crystal.

When researchers first began invading monkey brains in this way, they were trying to recreate H.M.'s lesion, the dragging cut through the medial temporal lobe. The lesion in his brain became known as an H + A + lesion, for hippocampus-amygdala, plus surrounding cortical tissue. In the 1970s, researchers at the National Institutes of Health, led by neuroscientist Mort Mishkin, tried variations on that lesion, looking to tease out which was the structure that ran the show. The NIH group reasoned that the hippocampus was part of a memory network, that other structures had to be involved. They focused on the amygdala, nut-shaped and still mysterious, close to the hippocampus.

As Zola-Morgan would do, the NIH scientists took a very direct approach, a series of surgeries to chisel the amygdala out of the brains of monkeys. The results were dramatic. Monkeys had a horrendous time remembering red blocks and blue balls. Suddenly, everyone was wondering if this little nut-shaped thing in the bottom of the brain was the real gatekeeper of memory.

Still, the picture wouldn't seem to stay in place. In rodents, where amygdala lesions were made, the memory of animals did not appear dramatically impaired. When human brains had extensive damage across that structure, the person's memory function was affected but not crippled. Yes, H.M. had lost that structure, but he'd lost a lot of other neurons as well. They couldn't figure out what else it could be. The slippery surrounding tissue had never been considered remarkable. The cortex was so much deadwood, according to the prevailing view of the time. Frustrated, Zola-Morgan, Amaral, and Larry Squire decided that, if they had to, they would rehash every monkey brain surgery done in the past decade. They would re-read reports until the words blurred. But they would find a pattern that fit.

Larry Squire, a lover of classical literature, remembers the turning point because of Shakespeare. He and his colleagues began seriously talking in mid-March 1987. Early in the tragedy *Julius Caesar*, a soothsayer warns Caesar to beware the Ides of March, the middle of that month as marked on the Roman calendar. Caesar ignores the warning, only to be murdered by a group of his political associates and friends. In their frustration, in their sense of reaching an impenetrable wall in the brain, Squire was fixing on gloom, thinking that the Ides of March was a lousy time to try to make a discovery, doubting that anything would come out of this time of foreboding.

The problem had started absorbing even their off hours. By the evening of Friday, March 11, the three scientists were glaring at each other, and getting nowhere as they compared notes. Where was the brain failing? They just couldn't seem to get it. On Saturday night, Squire and Zola-Morgan were on the telephone, still plugging. "Stuart and I talked for about an hour about the hippocampus, the amygdala and all at once we realized that we really didn't understand what was up there in the brain above those structures. Everyone had thought of the cortex as just a kind of cover. You cut into it getting to the hippocampus, but so what? The damage wasn't even mentioned in some of the papers."

On the Monday after their telephone conversation, Zola-Morgan and Squire met and began digging through slides of monkey brains. They headed over to Amaral's office. Amaral started putting the slides under his microscope. The room was completely silent, except for the whisk of slides in and out of place. He looked at three, four, five pictures of damaged brain. Then he looked up and said, " 'Every one of these animals has damage to the perirhinal cortex,' " Squire recalls. "I just threw my hands in the air and said 'That's it!' "

"There was just this memorable day," Zola-Morgan says. "Everything came together on that afternoon. There's this one moment in time when you've made a discovery, and you know that you are the only people in world that know. Think about James Watson and Francis Crick, and the moment when they deciphered the structure of DNA and they were the only ones in the world who had that picture, that famous double-twisted strand, in their minds. It just clicked. And the more we thought about it over the next few days, the more we cottoned to it, the more the idea grew, that it had to be the cortical regions, that we hadn't understood it, that no one had understood it, but that it just had to come together that way."

It has taken them almost six years to prove it, to convince the rest of the scientific community that those blank little blocks of cells are actually memory in motion. Amaral's painstaking brain mapping has made that undeniable.

One of the tricks of neuroanatomists is to inject into the brain materials that roll backwards from their entry point, retrograde tracers. With that technique you can shoot the hippocampus, for instance, full of color and watch it stream out, staining all the neuron connections into it, revealing the ties between the hippocampus and other parts of the brain. Today, the hot materials are dyes, once used to color fabrics, given names like fast blue and fluorogold. They reveal the pathways of nerves in bright colors, making the brain look like a road map. Red for interstate, a major rope of nerves through the brain; blue for state highway; green for two-lane back road.

Amaral's recent work, with graduate student Wendy Suzuki, shows the medial temporal lobe cortex to be almost blinding with gold and blue, messages flying back and forth across the brain's surface. There's an elaborate relay, crackling across the brain's surface. About two-thirds of the information from the perirhinal, for instance, feeds into the entorhinal. The entorhinal seems to sort it out, with most signals flowing into the hippocampus. They still don't know where all the signals come from, where they go.

It's the unknowns that make the surgery risky. They do not want to repeat mistakes of the past; they want to avoid slicing into the wrong places. But they can't be absolutely sure what the wrong places are. Zola-Morgan and Amaral have developed a means of using a needle-fine probe, threading it through the brain to the key structure and then sending a pulse of radio energy through it, destroying only whatever lies at its tip. If the entorhinal cortex is the target of the day, Amaral guides a probe to the narrow band of cells, brushes it gently with current, and then uses a glass pipette to suck away dead cells.

Slowly, it's becoming clear how vital those throw-away areas are: "We now think the perirhinal cortex is an absolute hotspot for visual recognition," says Amaral. The others are less clear, the perahippocampal cortex seems to handle a series of sensory inputs from other parts of the brain. Perhaps touch memory passes through here. The entorhinal may preserve spatial memory. And the magical little amygdala? Well, the amygdala is these days seen as a theatrical agent in the brain, a processor of emotion. Tucked into a nest of memory-processors, it is possible that the amygdala helps make remembrance into nostalgia, weaving emotions into memories.

The hippocampus, it is thought, may be a synthesizer, pulling the different aspects of memory together into one coherent picture. "We used to think the hippocampus did it all," Amaral marvels. "Now, we're trying to figure out exactly what it does do. And how to test that."

"Sometimes I can't believe how long it's taken," says Squire.

And on days like that one in January, when the monkey did die on the table, closing in on the answers seems like a dim prospect that stretches out into eternity. The surgery can appear flawless: into the brain, down into the entorhinal, back out, pack the muscle back into place, suture up the skin. All the time, the animal's heart beats steadily away, a reassuring rhythm of green peaks on the monitor screens. Then it slips away. Zola-Morgan spent ten hours in the operating room that day. The monkey never came out of anesthesia.

"We don't know exactly why it happened. We aren't positive how to prevent it the next time. I really don't like that," Zola-Morgan says. Maybe, he thinks, the death was the sum of bad angles. They were trying a sharper, cleaner dive to the base of the brain, pulling the animal's head back farther than in earlier surgeries. Perhaps the way the monkey's head was bent backwards choked the passage of blood between heart and brain. Perhaps the monkey was not as healthy as he seemed. They just don't know. But Zola-Morgan is not going to risk the angle again. He designed an air mattress. Now they rest the monkey's body on the yellow float, making head and heart more level, easing the flow. The next surgery went effortlessly. That animal was up and snacking by evening.

Eventually, usually within two years, Zola-Morgan will put the survivor under anesthesia again, and this time, kill it. He will open up the skull, remove the brain, and cut it apart. He needs to see the exact damage, to trace the probe. He needs to see how the brain may have reorganized itself to cope with loss. They can return to the brain more quickly that second time. After the first surgery, they do not replace the broken cheekbones, do not patch back the circles of skull they removed. They've discovered that packing muscle over the gaps is enough to protect a cage-bound monkey, that cheekbones are not necessary to a two-year survival.

The gaps are not painful, he emphasizes, and the broken bones are waxed, plugging leaks from blood vessels inside them. His experiments depend on animals that feel well and healthy, that can concentrate on the tests and tasks ahead of them. A good scientist takes good care of his animals, he thinks. It's a pragmatic approach, in a way, the approach of an intelligent user. A woodsman keeps his axe rust-free and sharp; a race-car driver fine-tunes her vehicle; a researcher, planning to ask lab monkeys to do demanding work, keeps them disease free, well-fed and rested, ready to go. The memory work done by the San Diego group is considered some of the best in the world. It is tough-minded, clearly thought-out science that continues to promise new discoveries about the human brain. Zola-Morgan knows that the intelligence of the monkeys makes it possible. He never fell into that trap, that Duane Rumbaugh describes, of considering the monkeys baloney-brains.

So it shocked him when animal activists protested the work. He was floored by their depth of anger over monkeys' deaths, over bloody surgeries, crippled brains. San Diego Animal Advocates have named Zola-Morgan "Vivisector (read "animal butcher") of the Year" three times running. They don't seem to get the point, or the promise, he thinks. They don't seem to catch the fascination or even to see the hope of saving lives: "It's so fundamental in a way," says Squire. "We were overlooking a critical part of the brain because we know so little about how it functions. We have to understand it to ever deal with diseases or damage there. Alzheimer's disease, when it impairs memory, begins in the hippocampus and entorhinal cortex. We understand that only because of this kind of work. And if treatments become available, we will know where to go in the brain. We won't be blundering around in the wrong place."

And yet, Zola-Morgan has run into absolute and sometimes terrifying hatred. He remembers ambling out of his basement office one day, into the dazzle of a San Diego afternoon, to watch an animal rights demonstration, a parade of placards and protestors across campus. Suddenly, he found himself staring at a demonstrator dressed up like Dr. Frankenstein's famous monster—green face, jagged black stitches and all. Around the monster's neck dangled a black-lettered sign that read: "Dr. Zola-Morbid: Give Me Your Brains and I'll Cut Them Up Alive." At another demonstration, a protestor spotted Zola-Morgan and suddenly flushed with fury.

"I was standing there talking to a colleague," he says. "And there was a line of university police separating us from the protestors. And all at once this guy just burst through the line, screaming, 'Nazi, murderer! Killer!'" Zola-Morgan is not a small man, standing six feet, but the man coming at him was bigger, he looked enormous, a Paul Bunyon of the animal rights brigade. "That was the one time I was really afraid. This guy was really big and he was crazy. He'd worked himself into this frenzy. I remember thinking, I'm going to get hurt here. But [the protestor's] friends were just grabbing for him and they pulled back. And police moved in between us. And I was very glad to have them there."

He thinks the activists must somehow romanticize what an animal is, imagine his caged monkeys staring through metal bars, remembering days of green and sun in the jungle. He knows they believe that the monkeys are born with the right to be outside, playing in trees. That they shouldn't be manipulating plastic cubes and hunting for M&Ms, waiting to be immobilized in operating rooms so that their skulls can be broken open. But monkeys, animals, don't have rights, Zola-Morgan argues, in the sense that people do. Human rights are a product of civilization, he says, a common agreement on acceptable behavior. Animals don't function at that level of sophistication.

"I think a human life is more valuable than an animal life. Humans come together with knowledge and a set of rules to live by, and a set of rights based upon those decisions. That's not a development that's occurred in the animal world. They don't have rights, they haven't developed that concept. That doesn't mean that we don't have responsibilities. We do. We have a real obligation to care for these animals well. But is my son's life worth more than that of a monkey? I don't even have to think about that answer."

He thought, at one point, that he could somehow share his own excitement, that he could say to protestors, as he does to his graduate students, "Imagine this black box." He thought he could make them empathize with H.M., lost in a warp of time. He thought he could bring them with him. Zola-Morgan has taken his case onto radio talk shows, into community meetings and public debates.

He thinks now he'd have a better chance of running into the tooth fairy one evening than of talking the animal rights community over to his viewpoint. If he tells them this is recognized as world-class science, that the federal government, after carefully reviewing his experiments, gives him $250,000 a year to pursue this work, that only makes things worse. They accuse him of ripping off the taxpayers just to beat up on monkeys.

Now, he believes that if he invited them to watch a surgery, they would see only broken cheekbones and bloody gauze. They would be unable to appreciate the gleaming beauty of the brain when it emerged. And he wouldn't invite them to watch, anymore than he would list his phone number or make his address public or even allow a newspaper photographer to take pictures of the locks on his cabinets. He wonders what's caused such people to lose their basic curiosity, the intensely human wish to understand, to open closed doors, to figure out why stars spit fire and why brains store such a wealth of information about something as insignificant as a chicken's egg. He wonders if his enemies haven't yet learned that all knowledge has a price, if they are simply impossibly and incurably naive. Are they without that most essentially human trait, curiosity?

"They're anti-knowledge and anti-progress," Zola-Morgan says. "And when the rest of our society really understands that, I believe we'll be able to turn this issue of animal rights back into the minor discussion that it ought to be."

FOUR

The Trap

THE LETTER CAME in a white envelope with no return address. Seymour Levine, a neuroscientist at Stanford University, studied it warily. He knew what it was; he'd gotten too many messages in plain wrappers already. They were never fan mail. He opened it slowly, reluctantly, a tiny rip of paper at a time. As if it might blow his fingers off.

There was a promise spelled out inside: "I resolve that unless you and your faculty recant your research and send the animals under your abuse to appropriate animal advocacy groups, my associates and I will respond with acts of physical retribution resulting in great bodily harm or even death to your family, your colleagues and yourself.

"Doubt neither our resolve nor our ability to realize this mission. Be aware that no channels of law enforcement available to you are sufficient to countermand our strike. We offer you no period of good will. You will not hear from us again. Do not force our hand. The loss will be entirely yours."

There was one other name in the letter, a sideswipe at the late Harry Harlow, a psychology researcher at the University of Wisconsin in Madison. "You and your sadistic father figure Harlow are as sick and unethical and bloodthirsty as anyone convicted in the Nuremburg trials," the letter said, referring to the trials of Nazi war criminals after World War II.

The message was signed by the Animal Liberation Action Foundation. The organization—if it ever existed—appeared out of nowhere in the spring of 1993, sending out a flurry of hate mail to animal researchers. Pharmaceutical companies received threats that products would be laced with poison. The Federal Bureau of Investigation stepped in. The conclusion was that it was an unusually nasty hoax, put on by some animal rights advocates who just wanted to scare a few, carefully targeted researchers.

Still, Levine started thinking he'd played human dartboard long enough. He was almost 70 years old. He'd spent his career studying animals—rats and monkeys—trying to understand the body chemistry of stress and of comfort. To do that, he had to first create the stress: induce fear, induce loneliness. He did it usually in the classic scientific method; separating rat families, taking young monkeys away from their mothers.

Over the years, he'd been able to painstakingly show not just that separation was traumatic, but that friends and family buffer an individual against emotional harm. With a support network, a monkey dealt better with everything—strange places, frightening predators. His detailed proof of that had helped promote support groups for people with frightening illnesses, like cancer, the deliberate reinforcement of social bonds. He was building a detailed picture of the body's response to companionship. He'd never thought of himself as a specialist in painful isolation, in the style of Harry Harlow.

Levine was tired of it, of the hate and of being afraid, of running a laboratory strung with barbed wire and electronic alarms. Some of the other letters had been less threatening, perhaps, but equally vicious. There was a determined ugliness. Another anonymous note ended this way:

"You and your associates are disgusting, loathsome degenerates. Probably you are a well-respected member of local society . . . Well, enjoy it while you may, for you will surely roast for eternity when they put your evil kike body to rest."

Maybe he'd just walk away from it. Maybe he'd just abandon experiments that he thought had real promise. He'd give up on his new rat studies, the ones that showed that a mother's touch was more than emotional comfort; it actually stimulated needed growth hormones. Instead, he would spend his time with his family. He sat talking, dreaming about it in the Palo Alto sunlight, in the small courtyard of his California laboratory. Behind him, dark evergreens formed a leafy wall, the air was squeaky with the high pitched chatter of 150 squirrel monkeys. Against the squeaks, Levine's voice rumbled, deep and quick. He balanced tensely in a white plastic chair, drinking coffee, lighting a ceaseless chain of cigarettes, letting the anger rise with the smoke.

"Don't I count in all of this? Maybe I've been dedicated enough. Maybe my wife is stressed enough. I'm not going to lie and say I haven't thought about quitting. People say to me, well, they win if you do that. And the answer is, don't I count? Yes, I think the work is interesting and valuable and it should continue. I think people who are doing it now should have a chance at their own opportunities. There are a lot of Stanford undergraduates interested in primate work and they have a right to do it. I don't keep it alive just because the animal rights people will win if I quit. I don't care about that, no, that's not true, I do care about that. But my feeling is, I've given at the office."

For the past decade, he's gotten hate mail from around the world. Protests are held outside his office. Threats come over the telephone. Complaints are filed against his work with members of Congress. By accident, by coincidence, by bad luck, he's pursued a line of research that is among the most despised by animal advocacy groups. If anyone wonders what kind of research can generate a decade's worth of hate, they need only to look at this laboratory. Stuart Zola-Morgan wonders where activists draw an unforgiving line: no intellectual curiosity past this point. The line runs here. For people dedicated to protecting animals, the knowledge Levine gathers is priced too high. Even the best of it, even the admittedly brilliant exploration of the linkage between mental and physical health. In a way, it doesn't matter how Levine pushes the future. He's trapped by the past. The rage against his work didn't begin with him, not at Stanford, not against the laboratories behind the evergreens and barbed wire. It began in Wisconsin, where Harlow rose to fame.

In his prime, Harlow was almost universally acclaimed. He was a scientific hero, one of those rare researchers who could charm both his colleagues and the general public. He was the author of more than 300 scientific papers and books, the winner of some of the most prestigious prizes in his field. Honored, admired, praised, and profiled, he received letters from young scientists around the country, asking for work in his laboratory. One of his star graduates, Stephen Suomi, now head of primate behavior programs at the NIH, wrote a dedication to Harlow on his Ph.D. dissertation. It read: "To the world he is a great psychologist. To me, he is a great man." Harlow's autobiographical accounts of his work were in such demand that they were published as pop-science paperbacks. In 1960, CBS TV's scientific series, *Conquest*, featured Harlow's work in an admiring tribute. This was after all the man who had redefined the bond between mother and child—an innovative researcher who had made love and affection subjects to take seriously in science.

In those glory days, it never seemed to occur to anyone to question how Harlow chose to explore the nature of love. It wasn't as if he kept it a secret. Far from it. His approach was very direct: the easiest way to investigate a loving heart is to break it; the shortest cut to explore a relationship is to sever it. After all, loss, grief, fear, stress are visible and measurable reactions. So, Harlow took infant monkeys away from their mothers and measured their response; he isolated them from other companions; he shut them, alone, into devices that would have trapped a Houdini. He kept them there for months.

Harlow's work made the importance of touch unforgettable. His studies convinced scientists that it's not enough for monkeys to see each other, they need—like people—to stroke, to hold. The science is known as maternal separation or deprivation or isolation. If Harlow wasn't the first to try it, he gets credit for building it into a profession, one that could be picked up and carried on by other scientists. Harlow made separation a scientifically respected line of research, solidly in the mainstream. Scientists like Levine are still conducting variations on those experiments, more sophisticated, more delicate, more savvy in their approach, but still testing the effects of broken bonds. Levine didn't work with Harlow, but in the sweeping anger of the animal rights movement, he didn't have to. He's carried on the tradition. It's enough.

It's curious, in a way, how unrelenting the animal advocacy movement is toward Harry Harlow, who died more than a decade ago, in 1981. In the period when Harlow's work flourished, researchers were doing far more gruesome things to animals than he ever dreamed up: In a 1957 experiment, scientists plunged unanesthetized rats into boiling water in order to measure blood changes; in 1960, scientists wanting to study muscle atrophy, immobilized the hind legs of cats with steel pins for 101 days, until the tissues withered. In 1961, researchers studied the effect of microwave blasts on dogs. Their detailed records noted that the unanesthetized animals began to pant rapidly as radiation increased, that their tongues swelled, that their skin crisped, and that if their body temperature was allowed to climb beyond 107 degrees Fahrenheit; the dogs died.

For that matter, monkeys didn't fare so well. They were heavily used by the military to test the effects of bomb-level radiation. In

1957, 58 rhesus macaques were put inside tubes set near the drop point—Ground Zero—for a nuclear bomb test. To no one's surprise, those set in tubes along the flashpoint of the explosion were fried. Still, those tests bred others. During the 1960s and 1970s, approximately 3,000 monkeys were exposed to up to 200 times the lethal dose of radiation, put on treadmills, forced to keep moving through electric shock, and measured for endurance. Army reports describe the monkeys as going into convulsions, stumbling, falling, vomiting, twisting in an apparently endless and futile search for a comfortable position. Researchers have shot rhesus macaques in the head with rifles, the barrel of the gun held just an inch from their skulls; shot them in the stomach with a cannon impactor accelerated to 70 miles an hour to study blunt abdominal trauma. Monkeys have been crippled by having weights dropped on their spines. They have been used in studies of organisms considered potential biological warfare agents; put into isolation rooms and sprayed with aerosol versions of the cholera bacterium, which kills through a ceaseless, draining diarrhea; injected with the Ebola virus, bringer of a hemorrhagic fever, which causes blood vessels to leak until a victim drowns in a rising tide of bloody fluid.

By such standards, you might think that Harlow wouldn't even make the blacklist. His monkeys were kept warm and dry. They were treated for illness, well fed, safely housed. They were not sliced open or poisoned or deliberately crippled. And yet, there are other kinds of pain. Loss, grief, and fear are destroyers in their own way. Using them deliberately means playing mindgames with an caged animal who doesn't understand the game. To animal activists, it's sadism, cloaked in jargon and scientific promises of aid to humans, but sadism just the same. There's just something cold-blooded about deliberately wrenching a baby from its mother. There's something cold-hearted about sitting for hours, methodically taking notes as it cries. ("Ooh-oooh-oooh," wrote one of Harlow's students, in a careful description of baby monkeys wailing for their mothers.) To most people, thankfully, radiation sickness is an unpleasant term on paper. But seeking comfort that does not come, a lost and frightened child crying for his mother—all of that is known human territory.

Harlow reported that—in the days before wide use of anesthesia—it sometimes took two lab workers to hold the struggling mother down while a third pulled the baby away. In an early experiment, students slipped a clear plastic shield between mother and baby monkey. The mothers raged, screaming at the researchers, pacing an angry path. The babies tried—as confused birds will smash into glass windows—to go right through the barrier. When it failed to give, the infants huddled against the plastic, as close to their mothers as they could get, and cooed to them—calling and calling—until the divider was removed.

That lost child image has haunted the field. Like Levine, other scientists who have carried on the field of maternal separation have been reviled and threatened by animal activists. This is particularly true for researchers who learned their craft working with Harlow. Among those are Gene Sackett, now at the University of Washington in Seattle, and William Mason, now at the California Regional Primate Research Center in Davis. Both men worked as postdoctoral researchers in Harlow's lab. A few years back, Sackett woke one morning to find his house spattered with garbage, pumpkins, ashes, the bodies of dead rats. Mason has been burned in effigy during protests in front of his laboratory.

Levine rages against the idea that he is guilty by association. He wasn't a member of the Wisconsin clan: he got his Ph.D. at New York University. He never even took a class from Harlow. When Harlow was making his name as a primate researcher, Levine was studying rats. He didn't even begin working with monkeys until the early 1970s.

"I was not a member of that club," he says. "The first time I met Harry Harlow, I was maybe 27 years old. I already had my Ph.D., I was working with rats at Michael Reese Hospital in Chicago. I applied for my first grant, and Harry came down as the site visitor from NIH. In those days, they had a one-person site visit. I was absolutely terrified. I don't think I've ever been more afraid. He was a big gun, very famous. He was editor of the most prestigious journal in our field." And did Harlow like Levine's work, a study of whether scientists affect the development of infant rats by handling them? "I got the grant, what can I tell you?"

It was that study that hooked him. The results suggested that a little stress can be a healthy thing, that the rats who were handled by researchers actually grew up healthier and more adaptable than those left strictly alone with their mothers. The conclusions were so surprising that, even today, researchers try to sort them out, hunt for the magic line between healthy stress and damage. Levine still pursues that goal himself, using blood-carried hormones as the meter of stress and change. It was a careful analysis of hormones in rats, for instance, that allowed Levine and colleagues at Stanford to show recently that when a rat mother licks her child, that touch turns on critical growth hormones in the infant body. The building of the baby's heart, for instance, can be disrupted if he doesn't receive enough maternal contact.

In squirrel monkeys, the species Levine prefers to study, the hormonal changes are also clear. From his view-point, the little South American animals are a wonderful test tube for stress; their bodies flood with the chemistry of alarm—a hormone called cortisol—when they are frightened or dismayed. Levine's interest has been in comparing cortisol levels in varying degrees of separation. He has taken infant monkeys away from their mothers; put them in total isolation, in a separate chamber where they could still see their mother, or simply removed the mother from the home cage, leaving the infant with familiar companions. The infants' stress response was lowest when left in familiar surroundings, greatest in social isolation.

In another study, in the early 1980s, Levine brought a caged snake into the laboratory, showing it to his monkeys, both in groups and alone. The snake was locked in a wire mesh box, placed near but not in the cages. Nevertheless, the monkeys, individually or in a group, were appalled. They squawked and chattered furiously, backing away when they saw it. Their body chemistry was radically different though; stress hormones soared in the lone monkeys. They rose just a little above normal in the monkeys with companions, even though they appeared equally alarmed. To Levine, it was an elegant proof of the stress-buffering effect of friendship. Every way he pushed that effect, it seemed to hold. He had also shown that companionship can calm a monkey against fear of electric shock. He doesn't consider the varied tests as redundant; just the building of good evidence in a case.

He's gone far beyond simple fear studies now, beyond basic relationships. Slowly, Levine has built a compelling case for the powerful connection between emotions and physical well-being, the scientific field of psychoneuroimmunology. The whole notion that stress makes us more vulnerable to illness, compromises our immune systems, is built on work like his. The link is being studied in young children, put in daycare, separated from their parents. It has already been put to work in cancer-patient support groups, at Stanford and elsewhere, acknowledging the importance of family and friends in dealing with terminal illness.

And yet, to Levine's frustration, none of that matters in his public image. The only aspects that animal activists concentrate on are the dark ones: that his past and his future hold experiments that, by necessity, may frighten or bewilder animals.

To Suzanne Roy, he's simply the man who uses taxpayers' money to scare monkeys with snakes. It might not be Harlow reincarnated, but it's too close. Roy serves as director of public affairs at In Defense of Animals, a tough and aggressive animal advocacy group. She's a dark-haired woman in her mid-thirties, a transplant from Washington, D.C. to California, with a political background and a tough, pragmatic outlook on running an anti-research campaign. She has been waging a campaign against Seymour Levine for three years now—partly, because she considers him a bad neighbor. In Defense of Animals also makes its home in the San Francisco Bay area, north and a little east of Stanford, in the bedroom community of San Rafael.

IDA's offices are deceptively modest, a storefront operation in a small strip shopping center just off I-80. The unplushy, bare-bones, warehouse kind of setting is deceptive, because IDA can be a formidable opponent. It's not huge; it boasts about 60,000 members and an annual budget of about $650,000. But it wastes little money on overhead (13 percent according to IRS records). Its strategy is pure pit bull. It picks a target carefully and refuses to let go. For several years now, IDA has been campaigning against research at the University of California, Berkeley, which it regards as basically anti-animal. It has been challenging military animal experiments, which it regards as vicious. Levine was picked with the same strategic motives. He embodies just about everything that In Defense of Animals dislikes about biomedical research.

"It's not the most invasive research in the world," concedes Roy. She enumerates the reasons the group zeroed in on Levine: the federal government gives him millions; he keeps expanding his use of monkey separation; it seems like a never-ending cycle, more money, more monkeys yanked from their family, more money, more monkeys. How many times does he have to prove his point, that monkeys feel safer in the company of friends or family? "It seems to touch a deep chord in people, taking innocent baby animals away from their mothers. You don't have to cut an animal open to torture it. And I think people understand that perfectly well. There's a strong emotional response. People are shocked. They say, I didn't know that was going on. And they say, why has this been going on so long?"

As part of its campaign against Levine, IDA printed up an angry little brochure in 1992. It reads, in part, "Like Harlow before him, Seymour Levine has come to symbolize the senselessness, waste, ludicrousness and cruelty of NIMH [National Institute of Mental Health]-funded research. . . . As long as Seymour Levine's work and reputation remains untarnished, graduate students will pass through his laboratory and go on to establish their own 'successful' careers in this cruel and wasteful field."

Harlow began his career at Wisconsin in 1930, after completing a Ph.D. in psychology at Stanford. He had every intention of working with rats. But when he arrived in Madison, he learned that the animal laboratory had been condemned and torn down. He camped in a basement office with as many mice and rats as he could cram in. He wasn't happy. As he told the story later, Harlow was grousing about his problems during a faculty bridge game one night. The wife of his department chairman suggested that he occupy himself at the local zoo in Vilas Park. She particularly recommended the monkeys, which she thought were charming and smart. At loose ends, Harlow ambled on down to the zoo.

He declared that it took only one monkey test to convince him that he never wanted to study a simple-minded rat again. The psychologist was fascinated by the idea of intelligence. He came from Stanford, after all, where Lewis Terman had developed the tests of human intelligence, the IQ tests. One of Harlow's early projects at Wisconsin was to adapt IQ testing to monkeys. The result, known as the Wisconsin General Testing Apparatus (WGTA), is still widely used today. The difference is in presentation. An intelligence test for humans may include a picture of several objects; the test-taker is asked to classify them, perhaps by geometry, squares versus circles. Monkeys can classify pictures as well, by studying slides. But the WGTA tests in a more physical way; providing a collection of wooden blocks, for instance, and asking the animal to sort them out.

The testing apparatus put Harlow's reputation on a fast track upward. But it was a cheese factory that really allowed him to start empire building. The university had acquired the abandoned factory on the southern edge of its campus—a gritty, industrial environment, far removed from the shining lakes and wooded hillsides of the central campus. But Harlow jumped at it. More than jumped. With his usual single-minded determination, he pitched in with the actual construction that transformed the factory into a lab. The rumor still runs at Wisconsin that Harlow would prowl around the campus at night, plucking choice lumber and other supplies from competing projects. The building, a squat rectangle of brick, painted the color of yellow mudflats, still houses the Harlow Psychology Laboratory.

It was in that boxy old cheese factory that Harlow stumbled into the isolation experiments. He was working with the popular research monkey of the day, rhesus macaques from Asia. His imported monkeys had been plagued by tuberculosis. He decided to buffer his pregnant mothers and their young by isolating them, first the mothers, then the babies. The first two such isolated babies were physically healthy. But mentally, they seemed somehow numbed. Even when placed later with other animals, they sat staring blindly through the bars as if alone. Harlow complained that they might as well be stone sphinxes. It worried him because he was also separating monkeys to study their intelligence. He wanted to chart their mental development, running them through his tests from infancy to adulthood. To do that, he was taking them away at birth. It was obviously not going well.

"Rhesus macaque mothers are not just going to say here, take my baby even once, much less over and over," says William Mason, who helped Harlow conduct some of the early intelligence studies. Mason is now a psychology researcher of national stature, the former president of the American Primatology Society. Like Harlow, Mason was quickly fascinated by questions of animal intelligence. The two scientists met when Harlow made a visit to his alma mater and they clicked. Harlow offered him the charge of a new research project and Mason leapt at it. "He was a very well-known psychologist," Mason recalls. "I knew I was being offered an exceptional opportunity." He arrived in the fall of 1954 and even the icy grip of the approaching Wisconsin winter didn't put him off. He dove right into the laboratory.

"They had six infant macaques who had been taken away from their mothers at birth and put into the nursery," Mason said. "As I remember, the people there had given them all 'stone' names, 'Millstone,' 'Grindstone,' probably because they had to get up and feed them at night." The nursery was efficient and sterile. Mason compared it to an orphanage of the time: clean, dry, plenty of food, minimal contact and affection. It became rapidly clear that something was missing from the formula. Instead of displaying the usual, chattering curiosity, these animals were huddling in their cages, wrapping themselves in the old cloth diapers or towels provided for warmth. In fact, the little animals, in the absence of a mother, developed a curious bond to the cloth. Some became so attached to the towels, they could be induced to perform the tests by taking the towel away and returning it as a reward.

Now, the idea of a surrogate mother was not a new one. Back in the days of Charles Darwin, biologist Alfred Russel Wallace, who almost beat Darwin to the theory of natural selection, had found himself the recipient of an orphaned baby orangutan. The little animal seemed to be constantly clutching for something, anything within reach, including, painfully, Wallace's beard. Wallace, trying to help, made a stuffed "mother" out of a roll of buffalo skin. He noted that the little orang clung happily to the dummy, becoming exasperated only when it tried to suckle. Still, most scientists thought of that as just a cute story. Harlow, observing the baby monkeys and their towels, began to wonder about surrogates. What made a mother important to an infant? Was it just something to hold on to? "There was actually a prevailing theory of the time that infants attached to mothers because they were the food source," Mason says. "No more than that. It sounds silly today, but back then it was taken very seriously."

Harlow told Mason to build some surrogate mothers. And, being Harlow, he had an eye on public appeal. So, he wanted the dummies to look like "mothers." He insisted on it. Mason, reasoning that the monkeys who bonded to towels didn't care about appearance, had first created a fat, soft roll of cloth, encasing a bottle. Harlow sent it back. He wanted a head. He sent it back again. The head, he insisted, was going to have a face. The monkeys, as it turned out, didn't care about doll-like heads—they routinely ripped them off or twisted them around.

"I don't think Harlow really liked animals," Mason says. "I don't think he hated them either. He just didn't have any feeling for them. He stumbled into animal research and he did it very well. He had a flair for sensing the currents of the time in psychology research. People were interested in the attachments between mothers and children. He sensed that this was the time to reach that audience, to talk about love. And early in his career, he had an enormous positive influence not only on the profession but the broader public. Not necessarily the man on the street, but educators, psychologists. He had a gift for spreading the word. People would listen to him and think, now here's a man who's on top of his field."

The question arises, would Harlow still be regarded that way, if he had held his studies in the gentle territory of simple affection?

What if he had been content to show that contact comfort, the ability to cuddle, was important to baby animals; that touch was an essential part of normal development? Or just to demonstrate that rhesus macaques were intensely social animals? In fact, as scientists looked at other monkeys, they realized that rhesus macaques were probably the most vulnerable to separations. Others, such as the squirrel monkeys studied by Levine, seemed to function naturally within looser social bonds. The little squirrel monkeys seem less distressed—less desperate, really—than members of the close-knit rhesus clans.

Harlow used rhesus macaques simply because they were available. "It was just luck to have started with rhesus, where the effects are so visible," says the University of Washington's Sackett, who worked with Harlow in the 1960s. Certainly, Harlow would have seen it that way, as luck. In the 1960s and the 1970s, people simply saw the animals more simply. They were tools; they were easy to use. Suomi's 1971 Ph.D. dissertation, done under Harlow, reflects the ethic of the day: "Experimentation . . . with human patients is seriously hampered by the lack of experimental control and sound ethical constraints. No such problem exists for the monkey researcher."

Harlow was a scientist, trained to ask questions and try to answer them. He took the question of affection and pushed it as hard as he could. What endeared the mother to the baby? He suspected that monkeys were bad mothers anyway. What if they were worse than indifferent? And what if she wasn't a cuddly, warm mother? What if she was cold—literally and figuratively? And what if she was abusive? Could you temporarily break the bond? And if the mother was abusive enough, could you smash that connection to bits?

As it turned out, you could create a perfectly vicious mother. But you could not drive the infant away, short of killing it. No one can fault Harlow for not trying. In the early 1960s, he had discovered that monkeys disliked being blasted with air. So he and his graduate students built a surrogate that blew blasts of pressurized air onto the clinging infant, whipping against the little animal so hard that the hair would be flattened on its body. They next created a mother who shook, rattling the baby back and forth until its teeth would literally chatter in its mouth. Another contained a catapult that would send a small monkey flying across the cage. And still another, Harlow nicknamed "the iron maiden". It was set with retractable brass spikes, which could be stabbed into the infant as it clung.

Harlow called these the evil mothers. He discovered that the little rhesus had an unswerving loyalty to them. Even with the iron maiden, even when the infants would skitter away, screaming with shock, they would stop, hesitate, watch until the spikes retracted, return to cling again. By putting females in isolation throughout infancy and then forcibly impregnating them, Harlow was able to create living evil mothers. The infants would come to cuddle. The mothers would slam them to the floor of the cage. Or kill them. One baby monkey starved to death; another was so badly battered that he was blinded and partially paralyzed. In one case a young mother crushed her baby's skull with her teeth. Another was found chewing the fingers off her child. And still the little monkey came back. This was its mother after all, and she—be she flesh and blood or wire and spikes—was all that baby had.

At the time, that notion of kinship, of born need, was considered appallingly unscientific—the stuff of sentiment rather than the stuff of research. In his own writings, Harlow talks about how difficult it was to even get a hearing on the concepts of affection and love, of bonding. He recalled one debate in which every time he used the word love, the other psychologist on the panel would carefully correct it. He was told that scientists didn't say love; they said proximity. Finally, Harlow snapped back: "It may be that proximity is all you know of love; I thank God I have not been so deprived." Harlow's willingness to challenge dogma of the time helped change how science—and the rest of us—look at mothers and children. Levine is no fan of Harlow. He considers him a grandstander and a troublemaker. But even he acknowledges the power of those first experiments: "His early work was superb."

The bond between young monkeys and their mothers is a sometimes troubling mirror of that between human children and abusive parents. The children still love their parents; they often blame themselves for the injuries—"My fault, I was bad." They may fear the parent, but they return. If you can look at that bond objectively, in the cold light of science, there's a compelling question there. Is there something in the basic biology of primate species—humans and rhesus macaques included—that weaves a bond tight enough to be deadly? So that a rhesus macaque will die in the care of an abusive mother—as some of Harlow's monkeys did—or a human child will die at the hands of its parent?

Compelling, maybe, but not an easy question and not a nice one. Rather, a deliberate move into darkness, a step into the twisted edges of normal relationships. You might think that a scientist, walking those edges, would want, at least, to discuss them with careful delicacy. It says a lot about Harlow that he did not. He was a scientist in the most unapologetic way: innovative, curious, fearless in moving onto new ground. Next, perhaps, he was a showman. He had a lively sense of humor. "He had this marvelous wit," says Mason. "A love of puns, and a real enjoyment of making his research so graphic and gripping that no one could turn away. What other researcher would call a laboratory device an 'iron maiden'?"

He was often a generous man; Suomi says he has never met a senior researcher so open-handed with his graduate students, so willing to share facilities and ideas. Another former student, Leonard Rosenblum, tells of how forbearing Harlow could be when plans went wrong. Rosenblum had decided to build a mechanical, threatening monkey head. He took Harlow up to the laboratory to see his device. Unfortunately, when he turned it on, a loose connection touched the metal framework. The head emitted a shower of sparks; the laboratory went dark. "Very impressive, Rosenblum," Harlow said, as he left the room. He never mentioned the disaster again.

And there were times when other scientists wished that Harry Harlow would just shut up. Classic Harlow was a 1974 interview with a reporter from the now-defunct *Pittsburgh Press-Roto*. Here's Harlow declaring his position on animal research: "The only thing I care about is whether the monkeys will turn out a property I can publish. I don't have any love for them. Never have. I don't really like animals. I despise cats. I hate dogs. How could you like monkeys?"

William Mason left the Wisconsin laboratory in 1960. He continued to do isolation and maternal separation experiments elsewhere. But Mason already had a sense of storm clouds rising. Harlow had continued to push the separation experiments. The laboratory built "isolation chambers," cages that screened monkeys from seeing others, faced them against walls. They left young monkeys in them, alone, for three months, six months, a year, two years. The two-year experiment was tried only once; the monkeys, reported Harlow, were mentally destroyed. Nothing the scientists did—pairing them with friendly companions, stroking them, giving them extra treats—could make them even lift their heads.

In the early 1960s, as abnormal monkeys began coming out of the isolation experiments, they seemed unable to reproduce. They were afraid of each other. The males and females would flee from any shared contact. Finally, the researchers decided that if the females were unable to flee—tied down, say—that maybe the males would have a shot at mounting them. So they developed a board with straps to lock the female in place.

Mason had done the early studies, showing the monkeys were socially, sexually incompetent. He left Wisconsin before Harlow tried tying the females down. But, of course, he heard about it.

"I'm not saying it was a good experiment," Mason says. "It was a crude effort to see if the males could inseminate the females. In retrospect, it probably wasn't justified. But, what really made it a problem, was that Harry insisted on calling this "the rape rack." Nothing neutral for him. He couldn't call it a restraining device. You can hear it now, and say he couldn't have picked a better way to offend. But that wasn't his intention, even. He couldn't resist the catchiness of it."

In his later career, Harlow seemed to develop a gift for offending women. Rape rack was only one example. He routinely referred to female monkeys as "the bitches" in his lectures. In one account of his observations of apes at the Vilas Park Zoo, he wrote of an aggressive female orangutan. The ape apparently settled down after the zookeeper, worried that she would attack a child, cracked a piece of wood over her head. The zookeeper, Harlow suggested, was obviously the man of her dreams, a person who understood the needs of women. In a 1973 interview with *Psychology Today*, he commented that he had been married twice, but "both my wives were too bright to be sucked into women's lib." He added that his second wife, Peggy, had lost her job in the psychology department when she married him, but "being a smart woman, she knew it was better to marry a man and lose a job than hold a job and not marry a man."

"Harlow gave a lecture on our campus toward the end of his career in the 1970s," says Gene Sackett, at the University of Washington in Seattle. "It was so insulting to women that, to this day, there are women in my department whose whole perception of Harry Harlow is based on that talk. You would have thought he would have toned it down. But he couldn't, I think. He had to be himself and he was a sexist. That was one of his problems. He was sharp. But he grew very rigid."

Harlow's former executive assistant, Helen LeRoy, insists that he was misunderstood on the issue of women. "I have the highest regard for him," she says. She argues that Harlow was supportive of women, that he talked to her, listened to her, never made her feel less than competent. "He was also a typical American male of his time," adds LeRoy, who still works at the Harlow Primate Laboratory. "I think the emphasis on women's lib in the early 1970s spurred Harlow into making cracks during speaking engagements just to stir things up. He did like to stir things up. These cracks went over like a lead balloon, of course. But they represent a very brief period in his career and public life."

Many people—both scientists and animal activists—have wondered if Harlow ever became unhappy with himself and his work. Some of his former students have suggested that, ironically, he became isolated by fame and acclaim, pressured to perform. That the public persona, the great and entertaining Harry Harlow, pushed him into a loneliness of his own. Other primatologists have suggested that, for all his bravado, Harlow was genuinely troubled by the darkness of his own work.

Toward the end of his career, Harlow developed a reputation as a alcoholic who was in over his head. Sackett recalls that one of his postdoctoral duties was to stop at a local bar each evening and to drive Harlow home, if his professor was unable. Levine, his own reputation growing, began to see Harlow frequently at professional meetings. "I put him to bed on a number of occasions, dead drunk," Levine says. "He was not a happy man. I recall when he became president of the American Psychological Association. He said, 'Now, I have nothing left to look forward to.' "

It was toward the end of Harlow's career that Mason, in dismay, began to think his former employer was sliding out of control, as if driven from within by a kind of evil spirit. Mason really likes monkeys; he's been known to take home ailing lab animals and cuddle them into health. "I know this will sound very dramatic," he says. "But it really was as if Harlow had his own personal demon. And the older he got, the more the demon was in control. He would write things about his experiments as if he did them with glee, as if he enjoyed the animal's suffering, that he couldn't wait to take these monkeys and destroy them. That's the sort of thing I got out of Harlow's later writings. They made my flesh creep."

Sackett recalls taking Harlow aside one day and saying, "maybe we should make this work sound a little less depressing." Harlow refused to even consider toning down his descriptions. "He said, you know I like to grab people's attention."

Their disagreement was over, perhaps, the most controversial experiment to come out of the Wisconsin laboratory, a device that Harlow insisted on calling "the pit of despair." The isolation studies, the ability to warp animals, had led Harlow to shift his direction. The monkeys were obviously troubled. Pacers, self-abusers, chronic masturbators, they were not normal monkeys. But they weren't suicidal either. Harlow pointed out that barring one monkey who died of self-induced anorexia, they still ate and drank. Harlow decided to take on that gap. It was, in a way, a search. Could you find some mental edge that separates anxiety, say, from depression and shove the monkeys over it? They tried first with a variation on the evil mothers. Knowing that the monkeys liked to cuddle up to their mothers for warmth during the night, they developed a series of mothers whose blood ran truly cold; it was ice water. The infants' responses were so severe that one died almost immediately; others curled themselves into corners, refusing to eat, becoming severely dehydrated.

Then another approach was tried, even more direct. If the temporary separation produced abnormal behavior, what would true isolation produce? He and his students set about designing a total isolation device, made to block out the rest of the world. The design they chose was a V-shaped metal chamber, with a mesh bottom, its sides polished to a slip-sliding gloss. A monkey at the bottom of the chamber could see nothing but the hands of a caretaker, who brought food and water. The animal was trapped in the point of the V. The results were immediate. When he wasn't calling the chambers "pits of despair," Harlow used another description: "the hell of loneliness."

Whatever that bleak line is—the borderland of depression, of misery—the pit successfully crossed it. In a way, the animals raised in the steel chambers had lost the ability to be monkeys. They were still alive, they still ate and drank. But beyond basic survival—a sea slug, after all, can eat and drink—the animals could not cope. They were not explorers, climbers, chatterers, fighters, or friends. They were dazed and unresponsive, in a sense like large, furry sea slugs. Some of them would finger the air, freeze in fright at the sight of their own arms—a characteristic of humans with severe schizophrenia. In a typical experiment, Harlow reported, the monkeys would spend the first few days scrambling around the bottom, slipping on the metal sides, trying to get out. They would spend the rest of their stay huddled, head down, on the bottom.

In 1970, Harlow decided to finish his career at the University of

Arizona and left Wisconsin. The remaining staff ripped out the vertical chambers; Suomi claimed that they gave him nightmares. He wasn't alone. By that time, too, there was a tide of rising public hostility against such experiments. Reaction was so strong that Sackett even today believes that the modern animal rights movement was, in part, born in the hissing anger over Harlow's laboratory. He blames Harlow for being too in love with flamboyant description to recognize that times were changing. That informing the public that you were cheerfully building "pits of despair" might be a major public relations mistake.

"The demon was in control," Mason says. "It wasn't as if they did that many extreme experiments there. The people I knew there were very concerned about the monkeys. But that gets covered up by the literature. You read Harlow's stuff and there's the 'pitiless' pit or whatever, and that's just a ridiculous thing to say. It sounds as if they were doing these experiments with glee. That Harlow enjoyed suffering, that was all this was about. Make these monkeys psychotic—take them and destroy them. They make him sound like the world's greatest sadist."

"He kept this going to the point where it was clear to many people that the work was really violating ordinary sensibilities, that anybody with respect for life or people would find this offensive. It's as if he sat down and said, I'm only going to be around another ten years. What I'd like to do, then, is leave a great big mess behind. If that was his aim, he did a perfect job."

Harlow's phrases dropped like plums into the laps of those unfriendly to animal research. Martin Stephens, a biologist and vice president for laboratory animal issues at the Humane Society of the United States, has analyzed Harlow's work on several levels. Of all the activists, Stephens has done the most scientific analysis of Harlow and his legacy, putting together a detailed critique of the field, sifting through more than 250 different studies. Stephens's 1986 report was commissioned by three other organizations, the American Anti-Vivisection Society, the National Anti-Vivisection Society, and the New England Anti-Vivisection Society. The result, nearly 100 pages of unfriendly commentary, was enough to annoy Gene Sackett into a formal, equally unfriendly review of Stephens in a scientific journal. Perhaps, says Stephens, the organizations wouldn't have paid so much attention if Harlow hadn't been so busy sending up signal flares—look at me, look at me.

"In a way, because of his eccentricity, Harlow invited criticism and attention," says Stephens. "More than any other psychologist, he was responsible for psychology being singled out for attention and focus by animal protection groups." God knows that Levine believes that. In an almost personal way, he blames Harlow for his own dilemma: "I'm going to get in trouble for saying this, but I remember a television show where they asked guests, what do you want most, fame or integrity? And when you look around at scientists with fame, it's been essentially at the cost of integrity. They overgeneralize; they overinterpret; they have an almost megalomaniac sense of their own importance. Harry Harlow is a good example of that."

It's true that the term "restraining device" lacks the emotional punch of "rape rack"; that "vertical chamber" is blander than "pit of despair." But bottom line, this is not an issue of language, not a quarrel over proper nouns and adjectives. It is a quarrel, often bitter, over what scientists do to animals; what they have right to do and whether there are limits on those rights. Is there a nice way to say, well, we take baby monkeys from their mothers, drop them into a steel chamber, and leave them there until we're sure they are drawn up into hopeless little balls of fur?

Stephens's criticisms of Harlow's line of work are threefold: that it is cruel; that it produced a dismaying lack of any results to help people; and that it goes on and on and on, torture without end.

More than forty years after Harlow began his experiments, Sackett and Mason and Levine are still pulling monkeys off their mothers, critics say. To animal advocates, to their public supporters, the scientists resemble a bunch of boys addicted to stripping wings off flies. Even worse, they get paid for it. If you hate the work, it's like a joke with a bad punch line: "Q: How many monkeys does it take to convince scientists that baby animals are stressed by being taken away from their mothers? A: As many monkeys as the government will buy for them." So far, more than 2,000 monkeys have been used in such experiments; some $60 million spent, mostly in federal funds. Seymour Levine at Stanford, for instance, is now in the middle of a five-year, $1 million plus grant, to study the stress induced by splitting apart squirrel monkey families. Suzanne Roy, at IDA, grudges him every dollar.

In the view of opponents such as Roy or Stephens, this is abuse of animals. The sum result of it has been—nothing to advance medicine or ease human suffering. Why, demands Roy, should the taxpayers be forced to subsidize middle-aged men who want to play mind games with monkeys? To hell with intellectual curiosity, is it right to tie an animal into mental knots because a researcher wonders if it would be more fearful of brass spikes or catapults? There's a frank demand in such questions. What, then, is the line that we are willing to draw, to say this knowledge is gained at too painful a price, with too much suffering?

If you want to study loneliness and isolation, Stephens and Roy point out, humans already provide plenty of material. There's, sadly, enough human loneliness and poverty and affliction in the world to learn that a barren and loveless home creates a troubled child. Further, they argue, there's little evidence that the work has done anything to actually alleviate the suffering of people. Roy has made a point of finding scientists critical of Harlow and Levine, and mailing their comments to members of Congress. One of them analyzed scientific reports on mental health issues, looking for any reference to the separation studies as important to the work. In all, Robert Bayard, a psychiatrist in Santa Clara county, examined 2,137 references in medical journals. He found no mention of animal isolation studies at all. "I conclude from all this that the animal infant-separation research has no value to anyone at all other than those engaged in the research itself. And I believe it is equally clearly not worth the suffering by the animals."

Stephens suggests that the greatest Harlow legacy to science is a generation of researchers who don't know how to do anything but redo aging experiments. "They make a career out of endless variations on a theme," he says. In fact, the animal protection movement is drawn to the isolation experiments because it regards them as an obvious example of waste in science. At In Defense of Animals, Suzanne Roy likes to compare Levine's work to the waste scandals in the defense industry—a few million on squirrel monkey experiments, a few million on pricey coffee pots.

In fact, says Stephens, the endurance of the primate separation experiments should serve as a clear illustration that science is a business, like any other endeavor. That scientists are in the business of raising money for their work. And if they find a profitable line of experiments, they will stick with it. "I'm not trying to discredit science," he says. "I'm saying in order to make a living, scientists have to mine some productive vein of experiments as long as necessary. I think, that instead of starting with human medical problems, the disciples of Harlow try to figure out how to best use animals and techniques they've been trained with. They want to use the primate colony next door. Sackett, at Washington, his job is to figure out how to make a living using those primates, not how to make life better for people with mental illness."

Certainly, the last time Sackett tried to do a classical isolation experiment, back in 1990, animal advocates dropped on him like waiting hawks. He had decided to apply primate isolation experiments to the problem of those mentally ill people who pick their own skin open, bang their heads on walls, eat glass, burn themselves deliberately. Science has a name for the syndrome, self-injurious behavior (SIB). There are some 160,000 people in the country who deliberately harm themselves in some way. Sackett knew well that isolation could push monkeys into hurting themselves. He thought the brain development and chemistry might be interesting in the self-abusers, reveal a pattern that might someday be applied to help people.

Sackett proposed to take 21 infants from their mothers and isolate them for a year with cloth surrogate mothers for company. Half an hour a day, they would be allowed to mingle with other monkeys. But that would be it. He expected that some of the monkeys would behave abnormally, perhaps chew on themselves. But they would have baby teeth, too soft and blunt to do any real damage. In rhesus macaques, the knife-like canine teeth don't usually develop until about $4^1/_2$ years of age. Sackett planned to remove the brains of the monkeys by 4 years, to study the effects of the isolation.

Before the experiment went to the National Institutes of Health, though, it was reviewed by the university's animal care committee. It was Sackett's bad luck that Seattle animal activists were attending that meeting in considerable number, planning to protest a rat head-injury experiment. When the monkey study came up, the scientists on the committee engaged in a lively debate. In the 1950s, Harlow's work might have gone unquestioned; in the 1990s, even researchers were uncomfortable with deliberately inducing animals to mutilate themselves. The experiment passed by a 10 to 7 vote. But members of the Progressive Animal Welfare Society of Seattle stomped away angrily, prepared for a full-scale attack. They raised money from around the country to hold protests and debates and to swamp members of Congress with letters of dismay.

The grant was not funded. Federal officials made a point of denying that they had been influenced by the furious Seattle campaign. Interestingly, neither side believes them. The activists have taken public credit for alerting members of Congress and generating enough political pressure to kill the project. Sackett thinks they are right; that the National Institutes of Health, the agency he applied to, was stopped by its own cowardice.

"It was a perfectly reasonable study and a study I would still like to do," he says. "But I believe even NIH was afraid to approve it. I think there are certain studies you just can't propose anymore. It's not so much that I'm afraid myself. But I can't risk my research assistants and everyone in my laboratory. I think it's quite possible that one of these far-out types will kill a researcher eventually. It takes a certain amount of bravery to do research at all anymore."

He has recently shifted his attention to less confrontational research, studying the stresses on female pigtail macaques, a more easygoing cousin of the rhesus, that may influence prenatal development. Sackett in particular is looking at how experiences during pregnancy may affect brain development. That shift can be seen in others who came from Harlow's laboratory. Stephen Suomi's primary emphasis these days is on the social needs of monkeys and how to respond to those needs in laboratory conditions. He has teamed up on such studies with Melinda Novak, at the University of Massachusetts, another Harlow graduate.

Mason has tended more toward trying to understand the animals themselves. His current experiments are with South American titi monkeys, tiny, fluff-headed animals, renowned for their enduring monogamy and strong family relationships. The titis, or callicebus, are sweet-natured monkeys, trusting in even their relationships with humans. Researchers can gently tug a baby monkey from a female titi and she relinquishes him; a female rhesus would do her best to take the scientist's hand off at the wrist. "The thing that fascinates me," Mason says, "is why creatures do what they do. When you say that the male and female callicebus have a pair bond, what do you mean. Is it emotional? Is it symmetrical? Does each animal feel the same way about each other? How do they feel about strangers, intruders? Is the male more jealous of the female, or is it the reverse? You can go on and on. But they're part of the overall picture. I want to know how these animals create their lives." His latest work sheds little insight on human problems, he concedes, except by simple kinship. That by understanding the way other primates construct their lives, we may better understand the evolutionary roots of our own behavior. Beyond that though, Mason argues that there is a place for straightforward fascination, for understanding better the other creatures who share the planet with us.

The titi monkeys have a 1990s kind of relationship, sharing the domestic duties. The father is the primary caretaker of infants. And in fact, Mason has found that the males are far more tolerant of a clinging baby than the females. In a series of separations, Mason was able to show that the mothers and fathers are more closely bonded to each other than to their child; that the mothers, when sequestered with the infant could get downright irritable, biting the baby's clinging hands or trying to rub it off against a cage wall, while the fathers were more relaxed. The infants seemed—not surprisingly—to be more upset by separation from father than mother.

This is, in fact, totally unlike the rhesus families, where the infants cling fiercely to their mothers. Indirectly, it reinforces another complaint of research critics: that you cannot use a monkey model of family relationships to learn about humans. Every species of monkey has its own social structure, each differing. You learn a different lesson in each case. And none of the lessons are exactly like those you learn from humans, who also have their unique social structure.

Mason concedes the point. In a professional journal, more than a decade ago, he raised that very question himself: "How far are we justified here in regarding monkeys and apes as useful substitutes for man? The hard fact—as we all know—is that, for most of the important psychological questions confronting man today, we search the crowded ranks of the non-human primates in vain for an acceptable human model. Our frailties, as well as our accomplishments, set us apart from the animal community."

Mason is unusual in his tendency to question his own work. He admits, laughingly, that he is rarely recruited to speak for animal research. He is a slim, bearded man with steady eyes and a soft voice. He is not a showman and, in perspective, he is about as far from Harlow as anyone can be in a similar line of research. He believes in his work. He believes that the research does help us better understand other primates, their role in the world, and our own. But he does not tell himself that he always acts in the best interests of the animals.

"I have never been able to reconcile fully the conflict between my interest in understanding the world as it is—that is to say, in science—and my feelings and moral attitudes toward the animals I work with. On the one hand, I believe implicitly in the value of the scientific approach. On the other hand, I believe just as firmly in the inherent value of the animals I work with and their right to a free and independent existence. Even if I were absolutely sure that my particular research would improve the human condition, I would find it very difficult to attach a specific moral value to my findings that could persuade me completely that the end justified the means."

In that, he stands in contrast not only to Harlow but Seymour Levine at Stanford. Levine is not troubled, as Mason is, by the complex relationship between scientist and laboratory animal: "I guess I feel that it's almost my fundamental responsibility as a scientist and a human being, if a problem needs to be solved, to try to solve it in whatever way I can. If using animals is the way, I'll solve it that way, just as people did when they wanted to solve the polio problem. The problems I'm dealing with may not be as directly, easily solvable. But that doesn't deny my responsibility to try. And I'll tell you honestly, the survival rate of my animals is a hell of a lot better in the lab than in the wild."

He doesn't kill animals in his laboratory; when they are no longer suitable for experiments, they go to other laboratories. The most physically invasive procedure he uses is taking blood samples out of the thigh, the blood-rich femoral vein. Those blood samples allow him to test for the stress hormones, such as cortisol. So, he sees it simply—he helps people; he takes good care of his animals.

"I'm not sure what the ethical issue is," he says. "I care for my animals. I really, genuinely like them. I like watching them and working with them. We are very careful to provide the best of all possible environments. They aren't isolated; they're in social groups, in large areas, with natural light. They get vitamins. And my monkeys either die of old age or are sent to another laboratory."

In a curious way, it's that sureness, superb confidence in his work, that has helped keep him linked to Harlow. Harlow, after all, rejoiced in strong language. He was no Bill Mason, with his quiet voice and ethical doubts. Levine also possesses a naturally flamboyant way of describing his work, an ability to attract attention. In this time of animal advocacy, there's a lightning rod aspect to his personality. "My wife says I'm bigger than life and maybe I am. I don't equivocate and I don't apologize."

Levine argues that he has reason to be sure of his work, perhaps more sure than the others. Coming up as he did, using rats for experiments, he developed an insistence on tightly controlling an experiment, on repeating it to make sure his results are no fluke, that his data is just so: "I had that drummed into me and the kinds of designs in my experiments are much more elegant than people without that background. [A lot of people in the field] don't have the same kind of controls. I follow up, I never publish just the first study."

Yet, for all his efforts to dissociate, there's no clean break either. Harlow's legacy won't be left behind, his influence on his field is unshakeable. And Levine has not avoided all of Harlow's goals; he just reaches for them in a more sophisticated way. In fact, Levine may have finally accomplished what Harlow vainly sought: creating a true biochemical model of depression. One of the stumbling blocks in isolating infant monkeys is that, although they may appear stressed, their body chemistry rapidly returns to normal. It does not mimic the chemistry of depression found in humans. A baby monkey taken away from its mother makes, at least, a chemical adjustment while still isolated. Scientists have found that if they show the infant a picture of his mother, his stress hormones shoot up again. But then he slowly adjusts again, the chemistry again leveling off.

Levine recently discovered, though, that if he takes adolescent squirrel monkeys—comparable to teenagers—and isolates them for several weeks, he gets a persistent chemical depression, remarkably like the imbalances found in severely depressed humans. The possibilities for testing drugs and other treatments, he thinks, are limitless.

In Defense of Animals is too smart to miss the connections with Harlow. It sends massive, carefully documented criticisms of Levine's work to Congress, just at the time when NIH's budget comes up for review. Members help organize formal protests at Stanford. The most recent was Mother's Day 1993, featuring a list of speakers mocking his studies of the parent-child relationship. IDA has organized letter-writing campaigns to Stanford, providing a basic form letter for complaining about his work. The organizers say they want to raise awareness. Levine blames them for stirring up hate as well. "You think those insidious bastards don't know what they're doing? Don't underestimate their cleverness."

The IDA brochure describes Levine as a disciple of Harlow. That makes him as angry as anything written there. "You know what disciple means. It means I was inspired by his work. And that's not true. It's dishonest. To them it's irrelevant though. Monkeys, infants, Harlow, put them all together and you're a disciple. And worse. You know, they say it like one word now:
'Harry-Harlow-infamous-monkey-abuser.' "

FIVE

The Face of Evil

Paul was a crab-eating macaque with a dragging left arm. The nerves from the spinal cord to the arm—the relay system from the brain—had been severed in an experiment, a study of the body's response to major nerve loss. Paul had been a chunky monkey once, weighing almost 20 pounds. But when he died, in 1989, he was down to a little over 7 pounds.

This is how he died: First, he began to chew apart his nerve-dead arm. Isolated macaques do mutilate themselves and Paul lived alone. He was too crippled, too defenseless, to be housed with another animal. The chewing could go on and on. In an arm without feeling, there would be no pain to stop it.

On February 16, 1989 he attacked the arm as if it was a snake, suddenly come to coil around him. His teeth cracked the bones in his hand. "His arm looked like it had been through a meat-grinder," says Marion Ratterree, a veterinarian at the Tulane Regional Primate Research Center, where Paul was housed. The vets decided to amputate at mid-arm, severing near the elbow. They were reluctant to take off the whole arm, which required breaking apart the shoulder socket.

After surgery, Paul went back to his cage. He refused to eat. His caretakers tried to comfort him, scratching his back. They tried to tempt him with peanut butter, rice cakes, sliced banana. He just turned away. He developed a wasting, draining diarrhea that responded to no drugs. He started ripping apart the stump of his arm again. Gangrene appeared in spreading black streaks. On July 4, Ratterree took off the rest of the arm, cracking apart the shoulder anyway.

Paul kept losing weight. They tried force-feeding him with tubes into his stomach, but he continued to wither. He lost the strength to stand. He died, down on the floor of his cage, head tucked against his remaining arm, on August 26.

You could say that any monkey put through a nerve-severing experiment is an unlucky animal. But Paul had the further misfortune of being more than an experimental animal. He was a symbol. He was one of the Silver Spring monkeys, removed by police from a Maryland laboratory in September 1981. Many believe that forcible removal marked the start of the current, combative animal rights movement.

Of course, the disagreements over research animals didn't begin at Silver Spring. There were too many confrontations before that. As Gene Sackett points out, animal activists were already angry over work like Harry Harlow's, already gathered in protest.

But Silver Spring began the most bitter 10 years in the history of relations between lab animal researchers and animal advocate groups. Before the 1980s, scientists did not fear activists so much, and activists gave researchers more benefit of the doubt. Now, they watch each other like wary enemies—each convinced that they are staring into the eyes of fanatics. Across that gap, negotiation and compromise can seem almost impossible. To understand where we are now—why the issue has become so polarized—you have to know what happened to the 17 monkeys from Silver Spring.

If Paul hadn't been one of those monkeys, he wouldn't have been caught in a cross-fire between researchers and animal rights activists. The vets treating him wouldn't have been forbidden by law to euthanize him. They wouldn't have had to watch him die slowly. There are few points on which animal advocates and scientists agree concerning the Silver Spring monkeys, but this is one: Paul died badly because the people on either side of this issue couldn't stand each other.

In addition to Paul, there was Billy, Sarah, Chester, Adidas, Big Boy, Augustus, Sisyphus, Haydn, Nero, Hard Times, Montaigne, Brooks, Allen, Titus, Charlie and Domitian. A 1993 *New Yorker* profile of the Silver Spring incident called them "arguably the most famous experimental animals in the history of science."

Certainly the story has been well-publicized: In May 1981, a 23-year-old college student, Alex Pacheco, applied for a job at a small private research center. The Institute for Behavioral Research was in Silver Spring, Maryland, prime science country, next door to the sprawling campus of the National Institutes of Health in Bethesda.

Pacheco interviewed with the institute's chief research scientist, Edward Taub. He claimed to be fascinated by research. Taub, then 50, was charmed by the young man's interest. He told Pacheco he had no money for a new position, but offered him a job as a volunteer. Pacheco took it. He did not tell Taub that his interest in animal research was not a friendly one.

He was already a seasoned animal rights activist. While in school at Ohio State University, Pacheco had organized such a scathing attack on local farmers, who castrated their animals without anesthetic, that a group of infuriated agriculture majors had camped outside his window one night, promising to perform the same procedure on him. After moving to Maryland, Pacheco had joined with another young believer in animal rights, Ingrid Newkirk, to form a tiny animal rights organization. They called it People for the Ethical Treatment of Animals. In 1980, it had fewer than 20 members.

Taub took Pacheco on faith. In an interview with the *Washington Post*, he recalled telling his wife, "I have a marvelous student. I told him there was no position, but he volunteered to work out of pure interest." Pacheco has always insisted that he picked Taub's building out of a hat, so to speak. He had obtained a list of federally funded research institutions using animals. The Silver Spring lab was closest to his home in Takoma Park, Maryland. He had no idea that Taub was in the business of surgically crippling monkeys.

The procedure Taub was using is known technically as deafferentation. It requires a surgeon to open the spinal cord and slice away the sensory nerves. The doctor could select the target, nerves leading to an arm, or both arms, or legs, or all limbs. Scientists had tried all versions, at times, numbing every limb in the body. They'd been experimenting with deafferentation since the 1890s, picking through the ways that nerves control the body. By the time Pacheco arrived, Taub had narrowed his focus considerably. He was interested in single-limb injuries. He had a group of sixteen crab-eating macaques and one rhesus macaque. Nine of the crab-eating macaques had been operated on; eight had lost all nerve connections to one arm. One, Billy, had both arms surgically numbed.

Taub wanted to challenge the long-standing theory that limb function was permanently lost when nerves were shredded away. He wanted to show that you could force recovery, perhaps even force new nerve growth. Force was an operative word. To make a crippled monkey use its bad limb, he tried strapping on a straitjacket to bind the good arm, leaving the animal only the damaged one. He tried putting the animals in restraining chairs and giving them electric shocks if they didn't move their numbed arms. He had shown that, under such duress, they could begin to regain function.

Taub believed that eventually such knowledge could help treat victims of strokes and accidents. NIH agreed. When Pacheco joined his lab, Taub had just received a $180,000 grant to continue his work. After a year, he planned to examine the spinal cords of the crippled macaques, to see if all that forcing of motion had stimulated new nerve growth. The procedure would necessitate killing the animals.

Pacheco didn't see innovative science and he didn't see promising medicine. He hated the place. He thought it was filthy, smelly, smeared with animal feces. There were roaches skittering through the cages (although one of Taub's defenders would later argue that the insects were a good protein source). The cages were rusty and old and small; each monkey was boxed in a space about 18 by 18 inches. Twisted bits of protruding wire had to be clipped off almost daily. Pacheco made notes of finding monkey corpses bundled into an old refrigerator, one body floating in a barrel of formaldehyde.

The survivors showed the classic stress symptoms of macaques caged alone. Some of the monkeys were spinning themselves around, banging off the cage walls, masturbating compulsively, chewing themselves open—especially the numbed arms. Their limbs were raw with bites and spreading infection.

Pacheco didn't bother to find out if Taub, if pressed, would have improved conditions. In his mind, he says, he had become an unofficial undercover investigator. He didn't want to blow the cover. With no one else in the lab bitching about conditions, he figured he'd just be fired if he complained. Further, he thought whistleblowing might have a more dramatic effect. "I was trying to clean up the whole system. If I'd gone to him, at best I might have cleaned up one lab and gotten myself fired." So, he began documenting the problems in secret. Pacheco told Taub he wanted to work at night. Still trusting, admiring his volunteer's dedication, Taub gave him the keys to the building.

In the dark, Pacheco would go inside to photograph, carrying a walkie-talkie. His PETA partner, Newkirk, would stand outside, ready to alert him if trouble arose. In late August, Pacheco began bringing in veterinarians and scientists known to be sympathetic to animal welfare issues. He asked for, and received, affidavits about conditions at the laboratory. The scientists were blistering: "I have never seen a laboratory as poorly maintained," wrote Geza Teleki, a primatologist at George Washington University. In early September, PETA took the photos and documents to the local police department. On September 11, Montgomery County police seized the monkeys. After reviewing the evidence, in particular, photographs of the monkeys, limbs—chewed open, bloody, oozing with infection—the police department filed animal cruelty charges against Taub, 17 counts, one for each monkey.

Seventeen charges, but to some biomedical researchers, they might as well have been seventeen rifle shots. They were a volley from the enemy; a declaration of war. For all its impact, though, it's important to keep Silver Spring in context. It was a turning point, beyond a doubt, but it was hardly the birth of animal welfare movements in this country.

There were animal activists busy on this continent before there was even a United States of America. In 1641, the Puritans of the Massachusetts Bay Colony drew up a list of liberties. Their formal legal code included "Liberty 92" which reads: "No man shall exercise any Tirranny or Crueltie towards any brute Creature which are usalie kept for man's use."

During the nineteenth century, as rising industrialization drove concerns for civil liberties and oppression, animal activism became more organized. The American Society for Prevention of Cruelty to Animals (ASPCA) was founded in 1866; the American Anti-Vivisection Society in 1883; The New England Anti-Vivisection Society was chartered in 1895. Vivisection means to cut up a live animal, emphasis on live. Dissection, by contrast, usually refers to cutting up a dead animal.

The rise of animal welfare organizations in the United States marched, almost in lockstep, with the rising use of animals in biomedical research. Laboratory use of animals rapidly became a prime target of the new animal welfare groups. The American Medical Association created its first formal committee to defend research in 1884. For most of its history, the American animal welfare movement has armed itself to take on the country's research community.

Science historian Susan Lederer, of Pennsylvania State University, points out even Pacheco's techniques at Silver Spring had been tried earlier. PETA was not the first group of activists to gain information by masquerade. In the 1920s, a group of animal activists in New York infiltrated a laboratory. They brought out photos of dogs with their mouths taped shut, so that the animals could neither eat or drink. The researcher involved—like Taub in the 1980s—was formally charged with animal cruelty.

Still, the passion of animal advocates for monkeys is recent, bred largely by our growing awareness of animal intelligence and the close relationship of primates to humans. The early fights were mostly about cuddly domestic animals, especially dogs. By the mid-1920s, anti-vivisection groups had been trying for decades to get laws passed controlling research on dogs. They could, and did, cite the worst of the current experiences: researchers who poured boiling water into the open bellies of dogs, crushed paws with pinchers, puppies' feet held in gas flames. It didn't matter. They were effectively countered by scientists arguing the need for medical research. Researchers insisted the dogs' injuries were necessary to understand pain, burn damage, tissue destruction. They couldn't heal, doctors said, without such knowledge. Those arguments were so widely accepted that, in 1926, the dean of Johns Hopkins School of Public Health declared the movement dead and its supporters "pathetic," Lederer reports.

Scientists made vicious fun of people opposing medical use of animals. Lederer has drawn together some graphic accounts of the times. The most combative example is probably a New York physician, James Warbasse. In 1910, Warbasse, an ardent supporter of animal research, began warning of the dire consequences of becoming too fond of dogs. He suggested that affection for the animals was actually a kind of psychosis, sexual in nature. In support of his argument, he cited the work of a German researcher who had divided women into two types, "the mother class and the prostitute type." Warbasse warned that women who liked to "fondle" dogs did not belong to the mother type.

Other scientists, Lederer says, were less antagonistic. But they made a point of warning their legislators that such activists saw dogs as the "wedge" into dismantling all animal research. Legislation to restrict research on dogs was introduced in Congress in the 1920s, 1930s, and 1940s, and was shot down every time. The power of the biomedical community—along with the public's faith in the goodness of science—was so potent that the United States didn't have a comprehensive animal welfare act until 1966.

If credit can be given to a single person—or animal protection group—for changing that picture, it goes to Christine Stevens, who founded the Animal Welfare Institute in 1951 and the Society for

Protective Animal Legislation in 1955. She still leads both organizations, headquartered in Washington, D.C., and has achieved a national reputation as both a strong and rational voice in the animal protection movement.

Stevens grew up with science. Her father, Robert Gesell, was an animal researcher, a physiologist at the University of Michigan. She was living near home, in the university town of Ann Arbor in the late 1940s, working as a volunteer at the Ann Arbor pet shelter. The veterinarian who ran the shelter was a passionate believer in animal research. So much so, that he was hiding dogs so that they could not be adopted, locking them in closets before driving them to laboratories.

"When we started investigating, we found claw marks on the closets," Stevens recalls. "The animals had been scratching to get out." Stevens knew from her father that the animals did not go onto to a life of ease in a research laboratory. She decided to try shutting the pipeline. It turned out to be a decision made at a remarkable time. The National Society for Medical Research (NSMR), a powerful lobbying group for medical scientists, had begun arguing for pound seizure laws, which would legally require animal shelters to turn their dogs and cats over to researchers.

In 1948, Minnesota passed such a law; in 1949, Wisconsin followed. Alarmed, Stevens went to the American Humane Association. By that time, though, animal welfare groups were weary of fighting science. Stevens asked the society to work against laws that required shelters to unquestioningly give animals to scientists. She offered $10,000 raised among friends and family to underwrite that effort. The AHA turned her down. In fact, an internal debate over taking on animal research eventually split the organization in two. The dissidents formed the Humane Society of the United States. Today, the American Humane Society is much tougher on research issues. (A recent national mailing strongly opposes the use of pound animals in research. It features a picture of a shaggy dog and this warning: "He loves people. He'll give you his paw if you ask him to. He's a 'perfect' laboratory specimen.")

Back then, it was different. "They were afraid of offending the medical people," Stevens recalls. "Medical researchers were riding very high. We couldn't even get a serious discussion going about what to do." In frustration, she founded the Animal Welfare Institute, determined to take on pound seizure laws herself.

Within a year, she received a hostile letter from an official at the

Federation of American Societies for Experimental Biology, warning her that her group, which had been perceived as "on the fence," was in danger of joining the "group of foolish, misguided, misanthropic, idle, mischievous or mercenary social perverts who form the so-called anti-vivisection societies to a large extent."

It's been 40 years and she can still remember how angry she was. She wasn't trying to stop research. She just didn't think researchers had a right to every animal, to scoop them by the handful out of animal shelters. She became convinced that scientists could not be trusted to regulate themselves; they were blinded by their own sense of mission. After achieving success against pound seizure laws, in 1960, she began working for a comprehensive animal welfare act that would, for the first time, include laboratory animals. The opposition from medical researchers, led by the National Institutes of Health, was fierce and absolute. So many representatives of the biomedical community testified against such legislation that, when it did pass, in 1966, Congress decided against an NIH request to do the lab inspections. Legislators feared the agency was too biased. They put the U.S. Department of Agriculture in charge of inspecting the laboratories.

Scientists did not welcome the USDA inspections and they did not rush to improve laboratory conditions. Stevens began to wonder if anything had changed since the law was passed. A born investigator, she began collecting USDA inspection reports. The collection turned into a list; the list became a book called *Beyond the Laboratory Door*, published by her institute.

The book is admittedly focused to the worst case, packed with photos of dogs blackened in burn experiments; cats with metal boxes fixed between their ears; monkeys with eyes surgically sliced out. The print section, the list of violations found by USDA inspectors, is almost equally grim: animals cages without food or water, cages slimy with feces, crawling with insects, spattered with blood. A single example: in 1980, the USDA brought a complaint against Ohio State University concerning its care of young cats in medical experiments. The complaint stated that "The kittens had lesions around their necks and many had metal tags with identification tags imbedded in the flesh of their necks. These injuries apparently resulted from chains being placed around the kittens' necks when they were young and not being replaced or lengthened as they grew older." Ohio State received what Stevens considers the lightest of slaps. It paid only a $500 fine even though the charges involved nearly 40 animals. The university also refused to actually admit to the charge of animal cruelty.

Furious, Stevens immediately began pushing for a strengthened Animal Welfare Act. And now, in the newly energized 1980s, she found strong support from groups such as the California-based Animal Legal Defense Fund and from the Humane Society of the United States. This time, too, members of Congress were running out of patience. The tougher standards gained the support of two powerful legislators, U.S. Senator Robert Dole, a ranking Republican from Kansas, and U.S. Representative George Brown, a Democrat from Southern California. Stevens notes that conservative Dole and liberal Brown made a truly odd couple and an undeniably forceful one. In hearings on the revised act, which became law in 1985, Brown invited her to testify before Congress, detailing some of the abuses she had found.

"Still, I think, animal protection groups had been hesitant to challenge the science establishment," she says. "That changed when Alex Pacheco went into the Silver Spring Laboratory. I'd been visiting laboratories, but I'd never seen anything as bad as that." Of course, she pointed out, her organization always announced its visits, while Pacheco's strategy was pure undercover.

In hindsight, Silver Spring was a warning flare from the animal movement: We're out of patience. Pacheco and Newkirk were not interested in polite, sissy tactics. History suggested to them—with some justification—that the system was weighted heavily toward researchers. They were ready for a fight and they were unafraid of confrontation. Before taking on Taub, in fact, Pacheco had signed on for a summer on the *Sea Shepherd*, a boat captained by an anti-whaling activist. It followed whaling ships on the oceans and rammed them. No wonder Pacheco didn't consider simply asking Taub to clean up the laboratory. He wanted the science exposed. He wanted it stopped. He wanted people to notice. He wanted them to be angry.

To the genuine shock of the science community, people *were* angry. Even with the undoubtedly devious aspects of what Pacheco had done, public sympathy swung to the crippled monkeys. Somewhere, researchers had lost that unquestioning allegiance. By the early 1990s, PETA claimed more than 400,000 members, a paid staff of over 100, an annual budget of nearly $10 million. There are now more than 400 animal advocacy groups in the United States, claiming a combined membership of 10 million and total income approaching $50 million. The money itself tells of the strength of support for animal advocacy; it comes, almost entirely, out of the pockets of American citizens unhappy about the treatment of animals. And often more than unhappy, genuinely angry. Out of the 1980s came a new breed of animal activists, fashioned in the image of Silver Spring—media-savvy, ruthless, and uncompromising.

Two cases in point: A veterinarian, Elliott Katz, heads the California-based group In Defense of Animals. IDA has been trying to end the work of Seymour Levine at Stanford. Its other local target is the University of California, Berkeley. Katz makes a point of holding UC-Berkeley's past against it. He cites—repeatedly—a worst case example from the 1980s, a dying monkey in a psychology experiment. The animal had been badly injured in a fight with another monkey. It was so poorly cared for that gangrene had set in. The sick animal had not even been provided with adequate food and water. The campus veterinarian, who euthanized the animal, noted that it had no water at all. It was drinking its own urine.

The university has spent millions improving its animal care since. It hired Roy Henrickson, formerly chief veterinarian at the California Regional Primate Research Center in Davis, to modernize its animal care. Henrickson came in determined to clean up the place. To Katz, though, Henrickson was "the pimp of the research community," and UC-Berkeley was forever untrustworthy in regard to animals. When Henrickson began overseeing construction of a new lab animal building, IDA repeatedly shut down the project with protests, at one point sending members to camp atop a construction crane. Katz's group sued for lab reports on all dead animals from Berkeley. It still distributes brochures describing that long-dead monkey. Relations are so tense now that Henrickson demanded his office be shielded by bullet-proof glass. "When Elliott first appeared, I tended to just think of him as a crazy," Henrickson says. "I discounted his ability to do harm. But I've learned not to underestimate the power of a fanatic."

A second case is in Los Angeles. The militant group Last Chance for Animals decided to take its quarrels directly to the homes of researchers. In 1989, attorneys for the University of California in Los Angeles obtained an injunction against the organization. LCA members had blocked neighborhood streets around researchers' homes, chanting, yelling, and banging on the scientists' doors, demanding that they come out, reveal themselves as murderers. In 1992, two members of the group were jailed for vandalizing a UCLA laboratory; they refused to accept a probation which would have barred them from protesting animal research. The tactic has caught on elsewhere. In Maryland, protestors have staged mock funerals outside the homes of scientists. In a fund-raising letter, LCA president Chris DeRose said "I've now gone to jail three times and I'm willing to do it again. . . . That's how strongly I feel about the atrocities committed against innocent animals in vivisection experiments."

The most feared and the most destructive of all such groups is the Animal Liberation Front. The ALF formed in 1982, a year after Silver Spring. Many scientists regard the front as a barely disguised division of PETA. Pacheco and Newkirk firmly deny any connection. The conclusion is based strictly on circumstantial evidence. Pacheco and Newkirk often praise the willingness of ALF raiders to risk personal safety for animals. PETA's literature describes the ALF as "the Army of the Kind." When front members conduct a raid—an estimated 80 such attacks have occurred since 1982—they leave it to PETA to publicize the action. Several state grand juries have investigated whether there is a link between the two groups.

The ALF has been destructive enough to gain a place on the FBI's terrorist list. But it is worth qualifying that description. Compare its activities to that of a no-holds-barred terrorist group such as the Irish Republican Army. The IRA runs deep underground. Members of the ALF also hide from law enforcement officers, but they also get chatty with the media. You won't find the bomb-planters of the IRA talking up their philosophy in celebrity news magazines. You will find one of the ALF founders—albeit wearing a mask—featured in a 1993 issue of *People* magazine, arguing for animal rights.

More than that, the IRA kills. So far—and despite the most dire warnings of the science community—the ALF has not done that. The group tends to prefer psychological warfare over the real thing. For instance, in 1984, two scientists at the California Regional Primate Research Center had ticking packages delivered to their doorsteps. The packages turned out to contain clocks and copies of Australian philosopher Peter Singer's 1975 book *Animal Liberation*, considered by many to be the bible of the current animal rights movement. The message was clear: We know where you live and we could hurt you if we wished.

They probably could. They are expert at break-in, vandalism, arson and theft—and not getting caught. The ALF "casualty list" compiled by biomedical institutions includes: 28 cats, used in nerve transmission experiments, stolen from Howard University in Washington, D.C.; 12 dogs involved in heart research, including 5 with experimental pacemakers, stolen from UCLA; $500,000 worth of damage to a laboratory at the City of Hope Research Institute in Duarte, California, involving the theft of 36 dogs used in cancer research, 11 cats, 12 rabbits, 28 mice, and 13 rats; 460 animals stolen from the University of California, Riverside; $50,000 worth of damage to labs at the University of Oregon; a diagnostic laboratory burned to the ground at UC-Davis, value of $4.5 million; 1,000 animals stolen from the University of Arizona where four buildings were broken into, two set on fire. The lab and office of a researcher at Texas Tech University Health Science Center were smashed and 5 cats stolen, along with records; the office of a Michigan State University mink researcher, containing more than a decade worth of records, burned to ash in a night-set fire. The list also reveals ALF's oddly split personality—lovers of animals, destroyers of property. Following a 1991 ALF raid at the Cook County Hospital in Chicago, the researchers returned to find that the midnight invaders had fed their baboons fresh fruit.

The ALF has not made the work of people like Christine Stevens easier. Scientists have publicly denounced all animal advocates as vandals and crazies, based largely on ALF activities. The illegal break-ins have overwhelmed the public's vision of the movement, too. Burning down a laboratory gets a lot more publicity than quietly lobbying for a new USDA regulation. Still, the label stings. Pacheco accuses researchers of a smear campaign: "There are millions of people in this country who are members of animal advocacy groups and to try to paint them all as terrorist is absurd." Most of the animal advocacy groups work the system: even PETA lobbies in Congress, even IDA sends carefully researched complaints to legislators. Stevens is not happy with the ALF. "I think they've gone overboard. And I think they don't make a whole lot of sense. Calling people in the middle of the night—that's not a wise use of your time. And they've gotten people seriously angry, which has worked against us."

Beyond technique, there is another crucial difference between Stevens and the fugitives of the ALF, one of the fundamental splits among animal activists: those who can live with animal experiments and those who cannot. Stevens is carefully neutral: "Our business is not to be philosophers," she says. "We take a purely pragmatic approach, that we try to do the best thing for the animals at the moment. They live a short time in research institutes. So if we don't seize the moment, we do nothing for them at all."

By contrast ALF—and PETA—seek an end to animal research. Pacheco points out that ALF members are merely vandals, while scientists spend their days mutilating thousands of animals. Balance it like that, and there is a curious logic to destruction. When a lab is trashed, research stops, at least briefly. In 1989, ALF destroyed the laboratory of neuroscientist John Orem, of the Texas Tech Health Sciences Center in Lubbock. His cats and his records vanished with the raid. Orem estimated that it took him a year to get his studies restarted. During that period, he was deluged by hate mail, receiving more than 4,000 letters. In a 1993 mail campaign, titled a National Referendum on Animal Rights, PETA asked recipients: "Do you feel that peaceful yet illegal activities are ever justified when their aim is the rescue of suffering animals?" The result was a strong vote of support.

And, to give the animal underground its due, there are instances when a laboratory break-in has done a great more than take a few rabbits out of the loop. Especially in the early days of the ALF, there were times when those raids changed the way science was done. The most compelling case—and undoubtedly the most influential—was the 1984 ALF break-in at the laboratory of Thomas Gennarelli, at the University of Pennsylvania.

Gennarelli was studying severe head injuries. He was trying to mimic whiplash in auto accidents, the jerking damage in which the brain is slammed against the side of the skull. He thought if the injuries could be understood in detail, they could be treated better. People might be saved. He had received about $1 million a year since the early 1970s toward finding better treatments for head-injury victims.

Gennarelli used baboons to explore the effects of head trauma, figuring that primates were the best animal model for human brain damage. He was a thorough scientist, meticulous about documentation. He videotaped everything: the handling of the animals, surgeries to explore battered brains, even the creation of the injuries themselves. There was nothing pretty about creating the damage. The animals' heads were encased in small plastic helmets, which later had to be pried or hammered off. A piston was slammed against the helmet in one pressurized blast, hard enough to pitch the baboons' heads backwards, causing their brains to skitter around, banging against the bony sides of the skull.

Someone—and it has never been revealed whom—from within the laboratory tipped activists off to the existence of those tapes. "Listen," says Pacheco, when asked. "It could have been anyone; the mailman from down the hall could have called it in. But, yeah, these raids usually start with someone from inside." Over the Memorial Day weekend in 1984, ALF raiders broke into Gennarelli's lab. They went straight for the videotapes, stealing 34 of them. Shortly thereafter, PETA started circulating an edited version of the tapes narrated by Newkirk and titled "Unnecessary Fuss." The title was taken from a comment by Gennarelli, before the raid, explaining that he didn't like to discuss his work publicly because people tended to overreact.

Pacheco edited more than 70 hours of tape down into the 20 minutes shown in the PETA video. Newkirk narrated in a voice soft and just short of sarcastic. It's difficult to put into words just how ugly that brief movie is. Within that 70 hours, there were 20 extremely nasty minutes. Even *Science* magazine, official journal of the American Association for the Advancement of Science, commented ruefully that "From a public relations standpoint, some scenes on the tapes—which were made for documenting the research, not for public viewing—range from embarrassing to disastrous."

Yet obviously, the laboratory workers weren't just "documenting" the research. Posing before the camera, young scientists held dazed baboons in silly "say cheese" poses; dangled them by crippled limbs, laughed when they struggled. Propping up one brain-damaged animal, whose paws quivered uncontrollably, researchers turned the camera on him and began a voice over: "Friends! Romans! Countrymen! (laughter) Look, he wants to shake hands. Come on. ... He says, 'You're gonna rescue me from this, aren't you? Aren't you?' "

Media-smart as ever, PETA ended the tape on that mocking plea for help. After strenuous public protest, including a sit-in at the National Institutes of Health, Gennarelli's funding was cut off. The government hit the university with a $4,000 civil fine after an investigation that included use of the stolen tapes. Today, PETA still shows the film on university campuses as a recruiting device. Today, Gennarelli is starting head-injury experiments again. But not with baboons. He's decided to use pigs.

Like Silver Spring, the Gennarelli break-in became more than a local incident. If Silver Spring had made it clear that animal activists would make their own rules, the cut-off of Gennarelli's funding made it clear that they could win, big. From that point on, people on both sides of the fence became more focused on winning the war. And they perceived it as just that, a war.

The Silver Spring raid and the Gennarelli tapes brought the plight of research monkeys into sudden, high profile. Primate researchers were shocked to realize they were no longer immune from attack. In the early 1980s, massive protest demonstrations were held at primate centers. They were the first of their kind: "Everything changed after that," says UC-Berkeley's Henrickson. "I used to look at my friends who studied dogs and cats, and thank God I worked with monkeys. No one cared about them. That turned around so fast it was unbelievable."

Yet, PETA had really only accelerated things. The pendulum had paused, ready for its return swing, as early as 1973 when a determined Englishwoman named Shirley McGreal decided to take on the cause of primates. Like many animal activists, McGreal's crusade began on a note of shock, confronted suddenly with trapped and miserable animals, and horrified. Like Christine Stevens finding the shelter dogs locked in a closet; like Alex Pacheco, remembering an early visit to a bloody, Midwestern slaughterhouse. McGreal's anger began in a very personal way. She was living in Thailand with her husband, John, an engineer. She arrived at the Bangkok airport, one summer evening at the same time that a shipment of young gibbons did. The little animals were dying in their cages. And when she looked around, she saw worse. The country seemed to her to be emptying its jungles of monkeys and apes, sending them—healthy or sick—to Western research institutions. She could hardly stand to go into the airport at all. By year's end she had founded the International Primate Protection League and, a short time later, pushed the Thai government into restricting all exports of gibbons.

McGreal has never been as single-minded as PETA, as focused on monkeys only as victims of research. From the beginning, her bent has been to conservation, critical of animal research partly because it strips endangered species from their forest homes. Hers is also the only animal advocacy group dedicated solely to primates, from lemurs to gorillas. By PETA standards, IPPL is a small group, some 12,000 members worldwide, an annual budget of about $250,000. But it is a lifework—you have only to visit McGreal's headquarters, in Summerville, South Carolina, to understand that.

Summerville is a small, oak-shaded southern town, inland from Charleston. McGreal lives on its edges, down a rutted dirt track that she formally petitioned to be named Primate Lane. Her combined home-office, a sturdy rectangle of brown brick, sits on a green stretch of farmland, surrounded by fruit trees gone wild and cages housing some 20 gibbons. The animals are refugees, former pets and former research animals. Gibbons are beautiful animals, graceful tree-swinging apes with impossibly long arms and small black faces fringed with fur. They swing themselves, hand-over-hand, so rapidly it seems one breath away from actual flight. But the most startling thing about gibbons is their sound. They speak in a vibrating, whistling call, as eerie and mournful as the echoing voice of loons across a Canadian lake. Gibbon calls are the only background music in McGreal's office, a strangely wild sound in the soft suburban greenery of South Carolina.

Like Stevens, McGreal is a believer in using laws and words. And like Stevens, she has found the combination effective. Many people credit her with helping to convince the governments of India and Bangladesh to ban the export of rhesus macaques in the late 1970s. She made a careful point of alerting both governments to the fact that the United States, contrary to formal agreements with both countries, was using macaques in military experiments—in particular, dosing them with bomb-level radiation and putting them through endurance tests. "We didn't intend to work for a ban in the beginning," she said. "We just asked the U.S. government to stop doing those experiments. But they wouldn't. So we had do something. We're an advocacy organization for primates," she says. "Animals need someone to speak up for them. So we try to take their point of view—would they want to be experimented on? And if you look at the human species, look at the whole picture, humans have done very little to justify the sacrifice of hundreds of thousands of monkeys. All life is precious—human life, animal life. But as far as I'm concerned, being a gibbon is more valuable to human life than being Adolf Hitler."

Unlike Stevens, McGreal is an outspoken opponent of animal research. She has joined with PETA in lawsuits seeking the custody of the Silver Spring monkeys. The issue in those custody fights, she says, is the same in all custody fights. It's about power. When the power was all with the research community, times were not so tense. But the balance has shifted. And in the aftermath of the Silver Spring raid—in the years of wrangling over the monkeys and watching them die—a new group of scientific activists has arisen, in some ways just as angry and unforgiving as their counterparts.

Edward Taub has never taken to the public warpath. He has defended himself when national journalists revive the issue. A 1993 *New Yorker* story showed him as a man who still somehow hopes to prove himself right; in one anecdote, he is laboriously bringing out photographic slide after slide of his defunct laboratory, trying unsuccessfully to convince the writer of his viewpoint. Taub's concern has been with his own perceived injustices. But he hasn't needed to tour the country, crusading against the animal rights movement. Others, far more willing to fight, have risen out of the ashes of the Silver Spring lab.

In fact, of all the scientists in the country who crusade for animal research, two of the most outspoken are Adrian Morrison and Peter Gerone. Gerone is an infectious disease specialist, formerly with the U.S. Army. He heads the Tulane Regional Primate Research Center, near New Orleans, where the injured monkeys were sent after Taub's laboratory was closed down. Morrison is a veterinarian and renowned sleep researcher, who holds joint appointments with the University of Pennsylvania and the National Institute for Mental Health. He was called by the federal government to defend the health of Taub's monkeys.

Morrison's public speaking style is soft-spoken and cool. But in private, he fires up. It is his words, perhaps better than anyone else's, that tell of alienation. Recalling a radio debate with Alex Pacheco in 1989—the summer Paul, the crab-eating macaque, died—Morrison describes it this way: "It was a debate through a telephone hook up. I did my end from my office at Penn. I spent 45 minutes listening to someone saying on the air that what I said was bullshit. That I was just a liar. After it was over, I just sat and sort of stared at the wall. And then I went down into the laboratory and I said, you know, I've spent 45 minutes looking into the face of evil. I'm not saying Pacheco is the devil or anything like that, but what was emanating was evil. And it's changed how I look at the world. As far as God goes, I'm still an agnostic. But I know there's a devil."

Morrison didn't meet Taub until after the cruelty charges were filed by police. But he was one of the few veterinarians in the country familiar with deafferentation. An old friend asked him to explain—and defend—Taub's work. After meeting with Taub, reviewing the information, Morrison and a colleague, University of Pennsylvania veterinarian Peter Hand, agreed to help. Morrison recalls that the night before the first trial, he met with Taub and his attorneys to review the evidence. "That was the first time I saw the pictures of the monkeys," he says. "I kind of gulped and said, okay, I'll still do it."

"I think the hardest moment of my professional career was sitting on that witness stand, looking up at the judge, and defending the way those monkeys looked," Morrison says. The pictures taken by Montgomery County police were dismayingly clear and crisp, unrelenting in detail. Five of the nine nerve-numbed animals had become self-mutilators; bones had been dislocated, surrounded by darkening swellings, open sores were draining from arm to hand. One of the monkeys, Nero, had apparently chewed off parts of every finger on one hand.

Still, Morrison and Hand insisted that this was a unavoidable consequence of the surgery. Macaques caged alone chew on themselves. They pick at open sores, and without pain, they picked constantly. They ripped at bandages, licked at salves. You had to live with nasty looking arms in such experiments. "Taub keeps getting judged by today's standards," Morrison says. "Today, we have different veterinary care. Today, I suspect, you couldn't even get those experiments approved because of the care requirements. But that was then." Further, the two vets pointed out that the monkeys were basically healthy. Their damaged arms might be an oozing mess, but they were chubby, bright-eyed monkeys. Morrison even today sees Taub as a victim, a scientist whose life was destroyed by a combination of malicious animal activists and bureaucrats afraid to challenge them publicly.

Not that every researcher defended Taub's laboratory. The director of one primate center pointed out that his facility, doing similar work, spent $2.50 a day caring for the animals. Taub's grant application requested only 50 cents a day. Even NIH publicly criticized Taub's care: sending a letter to *Neuroscience Newsletter*, saying that monkeys at NIH with similar injuries did not develop open sores when cared for properly.

Nevertheless, after two trials and several appeals, all charges of animal cruelty against Taub were dismissed. The toughest count had concerned one macaque named Nero. One of his numbed arms had become diseased and been amputated by NIH veterinarians. Eventually, it was decided that Taub could not be prosecuted for that injury because he was doing federal research within the normal bounds of science. The state of Maryland—as do the states of Colorado, Illinois, Missouri, Nebraska, North Carolina, Pennsylvania, and Rhode Island—specifically exempt federally registered research facilities from state animal control and cruelty statutes.

By that time, Morrison had become convinced that scientists were going to be on their own in a very tough fight. "Courageous administrator is an oxymoron," he says. "I remember at one hearing, Pete Hand turning to me and whispering "these guys are really scary." And he wasn't talking about PETA. He was talking about the bureaucrats. Here was a guy [Taub] whose life had been destroyed. And these guys were all diddling around, worrying about whether the i's had been dotted and the t's crossed. They get so caught up in the institution, it's as if they can't even see the people. And this is about people."

Even with his sympathies firmly with Taub, Morrison suspected that the monkeys were faring badly during the human battles. After being relocated several times—in a makeshift shelter in a volunteer's home, back to Taub's laboratory until one monkey died—the battered group of macaques had been sent to NIH's primate facility in Poolesville, Maryland.

Morrison and Hand went to inspect the animals there. The Poolesville facility was gleamingly clean, roach-free, many things that Taub's laboratory had not been. And yet, Morrison, on inspection visit there, experienced a sharp shock of sympathy for the embattled monkeys. "I turned to Pete and I said, 'Pete, if you were a monkey, which cages would you rather be in, Ed's or these?' " In the setup at Poolesville, the monkeys were in cages lined along a wall, facing a wall. At Taub's place, at least they could really see each other, they weren't turned toward concrete blocks. They were more together before, Morrison thought.

It might have been the closest Morrison and Pacheco ever came to agreeing on anything. Pacheco didn't like Poolesville either. He wanted the monkeys out of the clutches of NIH. The animals were already damaged enough by science, he reasoned. The experiments were over. Let the monkeys go to a refuge and be fussed over and coddled for the rest of their lives. PETA began a lobbying effort in Congress, to transfer the animals to a nonprofit wildlife sanctuary in San Antonio called Primarily Primates. Partly at Pacheco's urging, two legislators, Robert Smith, a Republican from New Hampshire, and Charles Rose, a Democrat from North Carolina, drafted a petition to NIH asking that the monkeys be sent to the sanctuary. Two hundred and fifty-two members of the House signed the petition.

If Pacheco had ever had even a flicker of faith in the science establishment, what happened next snuffed it out. First, the NIH director, James Wyngaarden, sent a letter of agreement to Congress, saying that he would allow the monkeys to be moved to a sanctuary. Wyngaarden gave another assurance, of great importance to PETA. He essentially acknowledged that the animals had done enough for science. In the letter, he promised that "these animals will not undergo invasive procedures for research purposes," that any analysis even of their bodies would be done only after they had died a natural death.

As it turned out, neither promise made in that letter was kept. Even at the time the letter was written, NIH was apparently trying to place the monkeys at a primate research facility under its control. Looking back, the politics of that decision are obvious. PETA was angry with NIH, but the biomedical research community was furious. In particular, researchers were outraged by the decision to close Gennarelli's laboratory; they thought NIH had crumbled under pressure. The Silver Spring case had been investigated by panels from the American Psychological Association, the Society for Neuroscience, and the American Physiological Society. All three had cleared Taub of any wrongdoing. The neuroscience society, in fact, voted to send Taub $5,000 toward his legal expenses. NIH was getting the message that it had better take care of its own.

Quietly, over a weekend in June, NIH associate director William Raub began negotiating with federally funded primate centers to take on the monkeys. He approached two of the most outspoken critics of animal rights—Frederick King, director of the Yerkes Regional Primate Research Center in Atlanta, and Peter Gerone at Tulane. Yerkes is operated by Emory University and, as King tells it, the administrators at Emory were not exactly thrilled to be pressured by NIH to take on a political firebomb. They agreed to take a couple monkeys. Then Raub pushed for a few more. Then a few more. "And then the administrators here just reached their limit," King says. "They said, okay, none of them. It was too much."

Down in Louisiana, Gerone was preparing to spend a quiet weekend, mostly sitting on the roof of his weekend cottage, nailing down shingles. Suddenly, he got a call from Raub. "NIH was really under the gun," Gerone says. "Raub wanted to know if we could take the monkeys by Monday." Gerone started hunting for Tulane administrators. He was lucky enough to track down Tulane's tough-minded president, Eamon Kelly. "I told him these were very controversial monkeys," he says. "That we were going to take a lot of flak. But that it was important that we hold the ground against the animal rights people. The first plan was that we would take the controls, and Emory would get the injured monkeys. But they said hell, no.

And then Kelly said, you know, taking some of these monkeys is like being a little bit pregnant. We'll take them all."

On June 23, 1986, the monkeys were moved to Tulane's primate center (then called Delta). The primate center is not at the university's main New Orleans campus, but across the long bridge that spans Lake Ponchartrain, in Covington, Louisiana. Covington looks like any rapidly expanding small American town, a strip along the main highway. In Covington, you see the Wal-mart, the Holiday Inn, the Burger King, in a gray-black sea of asphalt parking lots. The primate center, by contrast, is near invisible, down a narrow road, and deep into Louisiana woods, buildings and cages tucked among magnolias, pines, sweet gums, and oaks, shaggy with Spanish moss. Until the Silver Spring monkeys arrived, it was a quiet place. Within a week after the transfer, there were protesters lying in the entrance road.

"I knew it was going to be crazy," Gerone says. "But what I didn't realize was how personal it would be. That it would turn into, let's get Pete Gerone."

At first though, Pacheco was simply stunned. More than that. He felt suckered. He was suckered. "There were at least a dozen phone calls, from NIH lawyers to our lawyers, saying you'll have the monkeys next week. At one point, I even flew to San Antonio, bought the shell of a mobile home, and outfitted it as a clinic, so that there would be a clinic facility for the monkeys when they arrived. Because, you see, at the time I thought NIH was honest."

Instead, the monkeys were in rural Louisiana, far in miles and politics from PETA's Washington, D.C. organization. Further, they were sent to Pete Gerone, who had never been considered one of the great animal lovers of the primate community. Gerone had taken over management of the Tulane primate center in the early 1970s because he liked the scenery, he liked the idea of running a federal center, and he was impressed by the friendliness of Louisianans. His previous job had been at Fort Dietrich, the army's point lab for studying dangerous infections. The lab was like working in a big, impersonal vault. When he left Dietrich, Gerone recalls, his name was misspelled in the base newspaper. Gerone adds frankly that he didn't move to Tulane because he wanted to work with monkeys and that the animals haven't grown on him: "I don't relate to monkeys and I don't want to," he says. "It would bias my ability to direct a center that does research on them."

NIH had assured Congress that it had the monkeys in mind in

choosing Tulane; the animals would be able to live outdoors, enjoying the balmy Louisiana climate. Nevertheless, the crippled animals were never housed outside. They were caged in a rectangular, concrete block room. The starkness of their housing dismayed even some of the scientists who later toured the compound. Berkeley's Henrickson served on a panel of experts asked by NIH to assess the health of the monkeys. Robert Smith, the congressman from New Hampshire, accompanied the panel during its visit.

"And after the tour, Smith took me aside, and said, 'You know, these guys would make our lives a lot easier if they would treat those monkeys a little better,' " Henrickson recalled. "The setup was basic, steel cages in the middle of a block room. And I had to agree with him. If it had been me, even just for politics, I would have had murals on the walls, television, really fussed over those monkeys. I tried to take Pete aside afterward and suggest that to him. But he was so set by that point, he couldn't even hear me. He wouldn't have done anything that he thought might have made PETA happy."

PETA kept trying to get the monkeys away, locating other homes, enlisting congressional support. NIH rebuffed every suggestion; its position was not that it had lied. Administrators had simply changed their mind about what was best for the animals. And then, in July 1988, the agency suddenly became interested in the animals in a new way. Once again, the Silver Spring monkeys looked like promising experimental animals. Political circumstances—the fact that they had been kept alive so long—made them unique. Under Taub's original plan, they would have been dead years before. Now, they were the only animals of their kind, primates who had dragged around nerve-dead limbs for almost a decade. Neuroscientists, now focusing more and more on how the brain reorganizes itself after an injury, found the prospect of looking at the Silver Spring monkey brains irresistible.

The agency's administrators had to agree that it was compelling science. Trying to avoid a showdown, they worked out a careful plan. The animals would not be killed for the experiments. But when their ill health forced euthanasia, then scientists would induce a coma, open their skulls, and run some tests, looking for reactions in the brain. The NIH administration argued that this was not an invasive experiment, merely an exposing of the brain just before death.

It was not an easy sell. "That's a beauty isn't it," Pacheco says.

"We said to them, if it's so uninvasive how about trying it on your kids and wife?" The federal legislators were, perhaps, less sarcastic but equally angry. "I would consider it a very serious violation of a commitment to me, to the Congress, and to the public," wrote Carolina Democrat Charlie Rose in a blistering letter to the agency. He wasn't alone. NIH was flooded with angry letters as was the Bush administration; Barbara Bush alone received 46,000 pieces of mail. Pacheco was livid, furious with NIH, Tulane. "They are all liars," he says. "And after a while you realize that's the only way to deal with them, by recognizing that they will do nothing but lie and lie and lie. They have no interest in letting the public understand what's happening. All they care about is hiding the truth."

PETA went to court, obtaining an injunction to prohibit euthanasia as long as the brain experiments were part of the package. If the animals were incurably ill, Pacheco said, okay, then euthanize them. But don't make brain surgery part of the package. Once again, the research community perceived that PETA was trying to tell it how to do its job. NIH refused to back off the tests of monkey brain function. Peaceful discussions halted. "After all," Morrison says. "You don't negotiate when you are in a war." And, of course, that was when Paul began to die. Pacheco says he did not know that Paul was sick until his radio debate with Morrison. During that debate, Morrison blamed PETA publicly for forcing the middle-aged monkey to die a slow and painful death.

Pacheco immediately called PETA's lawyers, asking them to fax Gerone a what-the-hell-is-going-on letter.

"I say to Morrison, 'How the hell do you know this? You're not a party to the litigation.' So he says on the air, well, he's seen photographs. The monkey is starving to death, and PETA is making him suffer. Well. Tulane had been telling us that the animals were just fine. So, we faxed them a letter saying that, if it's true, they should put the monkey down immediately. And then we argued with them. Tulane doesn't want to do it right away. They want to wait; they want the experimenters to come down from Bethesda. We'd given them blanket permission to euthanize the monkey, but they wanted this court ruling instead. So it dragged out and the animal died. And then they say it's our fault."

At Tulane, the veterinarians caring for the monkeys, Marion Ratterree and Jim Blanchard, were beginning to feel betrayed. Pacheco had visited the facility in 1987 and 1988. He'd struck them as charming and rational. When the monkeys had first arrived, Brooks, one of the uninjured macaques, had suddenly died of pneumonia. They had feared a public trashing but it hadn't happened. There had been an amicable agreement to send five of the control monkeys to the San Diego Zoo. They thought they had a working relationship. Paul's death changed that. Sitting over coffee, sharing their frustrations, Ratterree and Blanchard came to a shared conclusion—that PETA wasn't interested in the animals as individuals, only as anti-science symbols. It was willing to drag out their deaths for the sake of a big public fuss. Ratterree began taking it personally: She hadn't been able to do her job; she'd had to watch an animal die; she blamed Alex Pacheco and his friends.

And Pacheco blamed her. He had come to see a conspiracy of "lies and excuses." His theory was that all the research community wanted was to win the fight. "It was all about the goddamn fight." In his vision, NIH and its flunkies wanted the monkeys dead. If they couldn't kill them overtly, they would kill them quietly, denying them a supportive environment, sealing them indoors, refusing them contact with each other, locking them in tiny cages where their muscles atrophied. He would listen to Ratterree's explanations: the monkeys couldn't live outside, maggots might infect the open wounds. He didn't really hear her though. He heard only the dark outlines of a plot.

Then Billy started to die.

Billy was the only monkey with two crippled arms. He moved himself around by scooting, bracing his weight on the backs of his numbed hands and rolling his body forward. Ratterree describes him as looking like a moving fur-ball. Because of constantly being bowed forward, Billy's spinal column began to curl. He developed pressure wounds on the backs of his hands, raw, red patches like bedsores. He rolled himself even farther forward. In a tight curl, whenever he urinated, he soaked himself. The uric acid caused a blistering diaper rash to spread across his lower body.

"Sometimes I think animal rights people don't know what it's like to deal with a wild animal," Ratterree says despairingly. "Macaques have got two-inch canines. They carry viruses that can kill you. So to treat the rash, we had to anesthetize him. And when you do that, they don't eat as well. They go off their food. And you can't feed them before the anesthesia, or they'll be sick. So we didn't want to anesthetize him every day. Sometimes, depending on Billy's mood, we could apply Desitin or A&D without the anesthesia. But one week, we had to knock Billy down five days in a row. We didn't want to keep doing that."

Then the monkey developed an infection in his bones. It responded only to one antibiotic, and that one tended to cause massive kidney damage. Billy had always been a relatively cheerful monkey, scooting himself determinedly around the cage. But he started showing signs of depression and weariness. He huddled into a corner. Ratterree and Blanchard did not want to relive Paul's death. They notified PETA that Billy was approaching time for euthanasia.

"And it was the same goddamned song and dance all over again," Pacheco says. "They wouldn't let us in. They'd let us have a second-hand report from a veterinarian. They wanted us to take them on faith. As if they'd given us a single reason to trust them." Gerone agreed to let PETA choose its own veterinarian to inspect Billy. Pacheco selected Tom Vice, a consulting vet with the Texas sanctuary, Primarily Primates. Vice flew in from San Antonio, examined Billy, and agreed that the monkey was miserable. He recommended euthanasia. He called PETA's lawyers to report. After consulting with Pacheco, the attorneys informed Vice that his recommendation was rejected: "We were standing here when he made the call," Ratterree says. "He was shocked. We were shocked."

But to Pacheco, this was just another monkey killing maneuver. He didn't believe Billy was that sick and they wouldn't let him in to find out. He started reviewing a list of lies in his head. They'd promised to cage Billy outside and hadn't, telling him that the monkey might be attacked by maggots. "Maggots," he snorts. "What horseshit." He didn't believe a word they said. He didn't believe they were applying ointment to the rash. He suspected that they might, sometimes, drizzle a little antiseptic powder on top of the monkey.

"I respected Tom's position. He's a veterinarian. He makes a clinical decision on when to euthanize an animal. But I told him I had to look at it differently. I had to look at it like this: If Billy was my brother, would I put him to sleep? Yes, if he was in misery. But if he wasn't really, if he had a chance of recovering, then I couldn't just let him die. I couldn't treat a member of my family like that. And I knew enough not to believe what they were saying at Tulane." He knew too that once Billy was under anesthesia, NIH experimenters would cut open his brain to study nerve connections. PETA might have gone for the euthanasia alone, but not—ever—in combination with the experiment.

On January 10, 1990, Tulane argued successfully before the U.S. District Court of Appeals for the right to euthanize Billy and conduct the neuroscience experiment. Billy was put down on January

14. That week, the Animal Liberation Front broke into the office of Adrian Morrison. Morrison had expanded his war on animal rights groups. He was actively helping other researchers under attack; he was writing letters in their defense. In the break-in, Morrison lost computer disks, videotapes, personal files, and an unpublished manuscript. ALF left him a message, spray painted on the walls: It read, in part, "For Taub."

Morrison was stunned. For six months, he debated whether or not to continue tackling the animal rights issue. "And I came out of that much stronger. My passion is to alleviate human misery," Morrison says. "It takes precedence over everything. And maybe I have a talent for this, for standing up for what I believe. If so, I should use it." He went public again, once again taking on the animal advocacy movement, PETA in particular.

By that time, PETA had filed a lawsuit for libel against Gerone. Gerone had become increasingly tired of seeing himself, his institution, and his employees trashed by PETA. Oh, there was an odd kind of glory to it: "If not for the Silver Spring monkeys, no one would have ever heard of me," Gerone says. "I'm known across the country because of this fight. I used to be very reserved. I've learned to speak my mind." The *Washingtonian* magazine had run a profile of PETA, implying that Pacheco had faked some of the worst photos out of Taub's laboratory. PETA sued the magazine, which issued a correction and apology. Gerone, impressed by the article, repeated its charges. PETA sued him as well. The suit was settled out of court. The terms are confidential. But Gerone has taken great pleasure in telling people that he was pleased with its terms.

"I was worried when they sued," he says. "But the university stood right behind me. People have told me how lucky I've been. The university has stayed behind me throughout every minute of this. I've been allowed to shout my position from the rooftops. If what I was doing had reflected badly on me among my peers or the university, then I would have put my tail between my legs. But when Alex Pacheco calls me a piece of dirt, that's just his style. I don't take him seriously."

For Gerone, Billy's death was something of a liberation. He believed that PETA's refusal to help a dying monkey exposed the group's real motives, as an organization that cared more about money and publicity than a small animal. They had handed him a weapon, he thought. And he used it, talking on radio shows, engaging in debates, mailing testy letters to people who wrote him criticizing his care of the monkeys. In one letter of response, he wrote: "What is really disgusting is how the Silver Spring monkeys have been exploited to fatten the coffers of the animal rights groups." Gerone admits he was starting to enjoy himself. "We had them [PETA] by the short hairs and twisting when Billy was sick and their own vet recommended he be put down and they said no," Gerone says. "I make good use of that one. I love to tell that story. So much for PETA's concern for the welfare of animals."

PETA and its allies continued to fight in court to keep the monkeys alive, and they continued to lose. The monkeys were euthanized in two sets: in mid-1990, Augustus, Domitian, and Big Boy were put down. In April 1991, although PETA fought it all the way to the U.S. Supreme Court, Allen and Titus were euthanized. All of them were put under anesthesia and scientists exposed their brains, testing them for signs of nerve reorganization. NIH neuroscientists found some strong evidence that the brains had tried to compensate for the injury, sprouted new nerve pathways. It was strong enough that they published a paper on their results in *Science* magazine, listing Edward Taub as one of the authors.

There are just two Silver Spring monkeys left at the Tulane primate center. Sarah, the one rhesus macaque, was a control monkey, her nerves left intact. She lives in an outdoor facility, playing grandmother to young, orphaned macaques. She has a calm, curious face and a direct gaze. The Tulane staff tried housing her with older monkeys, but she routinely beat them up. She's a good grandmother though, Gerone says. Nero is the last of the nerve-damaged monkeys. He has only one arm. He's the monkey who lost a limb to amputation in 1982. Nero lives alone.

Marion Ratterree describes Sarah and Nero as fat, contented animals, "two tubs of lard." Both are easygoing monkeys, by macaque standards, relaxed and accustomed to life at Tulane. Gerone insists that the animals could never have gotten better care elsewhere. The center has been very careful, guarding against any criticism that it did not maintain the health of the animals. "We even counted monkey biscuits. We didn't want anyone accusing us of starving the animals." Gerone has hopes that when Nero begins to decline, Pacheco will have lost interest in a losing battle. That he will just accept the animal has to die and that experimenters may be able to coax a little more information out of the brain of the last of the Silver Spring monkeys.

To that, Pacheco has a short answer: "Never."

SIX

The Peg-leg Pig

In the last of a gray December day, along the oak-shadowed edges of the St. John's River, Peter Gerone sat watching a slow nightfall.

There's a shimmer to the close of day in Louisiana, smoky with dusk and the silky, clouding humidity, almost soft to the touch. Gerone began relaxing into it. He was at Benedict's, one of his favorite restaurants, tucked between the straight main road through Mandeville, Louisiana, and the river. Gerone and his wife, Lois, had been married in this restaurant, a converted turn-of-the-century home with a curved central staircase, gleaming maple floors and wide windows filled by the shifting light of the river.

He had all the elements of relaxation there, the easy night, the glimmer of candles, wine in a crystal glass, a platter of spicy Louisiana oysters wrapped in spinach and cheese. He was maybe ten miles down river from the primate center, where the land also bumps against the St. John's River. He hadn't really left it though. For all the polish of the room and the night, his thoughts were still there, dwelling on monkeys—and the enemy.

One thought followed another, out of relaxation and into exasperation: the idiocy of animal activists, the bad influence of Disney cartoons, and the gullibility of the Bambi-loving American public.

"Walt Disney did us in," he grumbled. "Before all those movies, people looked at animals differently." He snorted with disgust, mulling over that trend of animated movies populated by big-eyed deer and big-bad hunters that terrify defenseless animals. Bambi, Thumper, Flower the Skunk, Gerone sees them all as trouble, part of the reason that we live in a country filled with people who seem only to see animals as cuddly.

He put down his wine glass, watching the clear liquid slosh against the edges. Before this century, he points out, most Americans lived on farms. They butchered hogs themselves, took chicken eggs, milked cows, ate the animals that surrounded them. They shot wild animals to protect their herds, to add to their food supplies. Now, the country's population has concentrated in cities. Hunting is largely a recreational sport; farmers are in decline. People are surrounded instead by pets, sleepy cats, playful dogs, pet rats and guinea pigs. "There are kids out there who think meat is born cellophane wrapped," complained Gerone. How can they identify with the idea that animals—the ones they play and feed and sleep with—should be available as tools, for research?

"The animal rights people are going after that generation, the urban kids," he warned. "They've infiltrated everything. You know, I got a bunch of letters from kids at one of the New Orleans schools, asking about the Silver Spring monkeys, asking why I tortured animals. I tracked down one boy who had an unusual last name. It turned out their teacher had assigned them to write those letters as an exercise in writing a business letter. Well, I called the principal and demanded equal time and you'd better believe he put a stop to it." Still irate over the memory, he tipped the wine glass again.

He belongs to a group of scientists, you might consider him a founding member, who are determined to be heard in the debate over use of animals in research. No more being shouted down by picketers, no more being terrorized by vandals. Gerone and his allies believe that there's more to fear in being silent. Unless they make their stand—pushing back when they are pushed, tracking down junior high school teachers who tell students that animal research is bad, organizing the resistance—unless they dig in, they fear that they will lose the best of their profession, the ability to pursue interesting questions unfettered. Even as Silver Spring and the rise of groups like PETA transformed the animal activism movement, so have the Pete Gerones begun to transform American science itself, giving it a new edge. They have their own mission, to make their profession battle ready.

At 65, Gerone remains slim, clear-voiced, exuberantly opinionated. He came to the Tulane primate center in 1971 from the U.S. Army's disease research center in Fort Dietrich, famed for its emphasis on biological warfare. Gerone was himself an infectious disease specialist, head of a division that screened viruses for useful weapons, analyzed viral transmission. One of his favorite projects involved experimenting on humans, a study of cold viruses using volunteers from federal prisons. The prisoners were placed in rooms with nozzles set in the walls and sprayed with viruses. The scientists then studied the spread of disease from man to man. They were meticulous in the way that only data-obsessed researchers can be, right down to weighing each prisoner's sodden paper tissues. "You can't do that anymore, but they were great studies," Gerone says. "The prisoners loved it. They got to lie around, eat candy bars. All they had to do was drop their tissues in bags for weighing."

But in the early 1970s, President Nixon was clearing house at Dietrich, dismantling the biological warfare programs. Gerone heard about the primate center job in Louisiana and decided to try for it. The rural atmosphere, the friendliness, the buildings set in the woods, all appealed to him. "I didn't know a thing about monkeys when I came here," he says. "I couldn't have told a macaque from a baboon." He knows them better now. Tulane runs the largest of the federal primate centers, housing nearly 4,000 rhesus macaques, crab-eating macaques, mangabeys, and African green monkeys. Their company hasn't changed Gerone's perspective, though. He has grown more familiar with but not more fond of monkeys. He sees no reason why standards for their care should exceed those for humans, including human prisoners.

With some impatience, he finds himself supervising scientists who get friendly with their animals. He can tolerate it, but it irritates him. One of his researchers, a dark-haired, dark-eyed Virginia biologist named Robert Gormus, has been studying leprosy using a group of monkeys from Africa, black-faced sooty mangabeys. Although Gerone discourages it, Gormus made a friend of one of his monkeys, a female that he has nicknamed Mrs. Mangabey. The small monkey has gained a center-wide reputation as a hugger, reaching to be picked up, cuddling people who pick her up. Mrs. Mangabey has even hugged Gerone's wife. Not the director himself. "I don't let monkeys hug me," Gerone says firmly. "It's not good to let them get too close. Bob Gormus is an animal lover. And I need to ask him what he's doing with that monkey. He hasn't used her in research yet, and we're not here to provide pets."

The lobby of the Tulane primate center is decorated with glass-fronted cases filled with skulls: rhesus macaque, crab-eating macaque, stumptail macaque, baboon, patas monkey, owl monkey, squirrel monkey, four chimp skulls clustered around one human skull. The bones shine dull ivory, fragile, different. The most obvious contrast between the human and the monkey is not in the size of the skull. The more startling one is in the teeth. Human teeth, exposed in bone, are the teeth of a species that has knives to help cut meat. They look domesticated. Monkeys have the teeth of wild animals, dagger-like canines and incisors, weaponry. That humans dominate their fellow primates has nothing to do with the nature of tooth and claw. It's in that sudden flaring of skull, the bone that wraps around the brain. The real difference isn't told in bone at all; it's what's inside. Gerone likes to emphasize both those differences, in intelligence and in civilization. It's why he's here, dining at Benedict's and why his monkeys wait in their cages, back in the primate center. Monkeys are not his friends and they are not his equal. It's a point he likes to make clearly. There's a joke—not about monkeys—he tells often, when he gives public speeches, to illustrate his point.

Over dinner, he gleefully recounts it: "There once was an animal rights activist who worked in Washington, D.C. She lived pretty far out in Maryland and to come into work everyday she had to pass through some fairly rural areas. One day, while she was driving into work, she passed a farm with a pig with a wooden peg-leg hobbling around in its yard. "Wow," thought the woman. "Now, there's a farmer who really cares about his animals. One day, I'm going to stop and ask him about that pig."

And one afternoon, when she left the office early, she did stop at the farm and the farmer was more than glad to tell her about the pig. "This a wonderful animal," the man said. "One night our house caught on fire. It was burning away and smoke was starting to come out the windows, but we were all sound asleep and we would have burned up except that the pig came and banged on the door and woke us up."

"And the pig's leg was burned in the fire?" asked the woman.

"No, no," said the farmer. "But let me tell you another story about this pig. One day I was driving my tractor and it turned over. I was pinned underneath and gasoline was pouring out. I called and called, and no one heard me except the pig. And it came running and pulled me out from underneath the tractor."

"And the pig's leg was injured?" said the woman breathlessly.

"No, no," said the farmer.

"But what about the pig's leg?" the woman asked.

"Well, that's what I've been trying to tell you," the farmer said. "This is too good a pig to eat at once."

Gerone tells the joke with flare, laughing on the punch line. The punch line is really not that funny to him; it's real. It goes back to his complaint, that Americans are too out of touch with hunter and farmer, with recognition that animals are here to help provide for our survival. He wants people to think about what the tradeoff is at its most fundamental—human life versus animal life. "In my own mind, it comes down to the question of which do I want to help the most, animals or people? And for me, that's a very easy question to answer."

Who comes first—the farmer or the pig? Or Gerone likes to make other comparisons: If your house was on fire, would you rescue your cat first or your wife? If you were standing by the edge of a pond, a baby drowning at one end, a rat at another, which way would you run? Is it such a bad thing to put one's own species first? Other animal species do. They use each other, right down to ants, which herd tiny aphids around, stroking the little insects to "milk" a sweet fluid that oozes from their backs on touch. Animals prey on each other as well. "The fox doesn't think about animal rights when it eats a chicken," Gerone says.

That doesn't mean doubts never creep in. "In my younger days, I never thought about it at all. Animals, as far as I was concerned, were put on earth for human use. But once someone raises the philosophical question, once I really think about it, once I ask myself just exactly what is it that makes me feel I have a right to use animals, I'm at a loss for an answer. At the same time, my feelings are not so mixed that I hesitate to use animals for the benefit of mankind. I've never paused—it would be dishonest to say so."

Animal advocates accuse researchers, often, of hanging onto experiments because there's money in it. Gerone doesn't deny that it's a factor: "I'm willing to admit that. I can't deny scientists are very strongly influenced by where the dollar is coming from." And there are large dollars at stake here: NIH alone allocates more than $40 million a year to its primate programs. Over all, experts say that about half the NIH's research grants, totaling near $5 billion a year, involve some animal research.

"But, bottom line, no one up in Washington, D.C. ever says, well, let's keep this project alive because Pete Gerone wants the money," Gerone says. The sum total of research grants does not tell the entire story either. Tulane's annual budget is close to $8 million, mostly from federal grants, but the individual researcher's actual income is rarely the stuff of dreams. In a low-budget state like Louisiana, primate researchers can pull in less than $30,000 a year. At Tulane, you go into primate research because it's a living, not a fortune. Or because, like Gerone, you believe that there is good medicine yet in working with monkeys and that its promise must be defended—vigilantly—against those who would take it away.

In his mind, the battle lines split cleanly: us and them, we and they. They on the opposite side, the dangerous side, the wrong side. "It's a war of philosophy," Gerone says. "And there's very little room for compromise. They want animal research stopped, and there's no way we can go along with that. If we win—and I think we will—it will be because more of us are willing to stand up."

If anyone had doubted that Gerone had learned to fight tough, he proved them wrong in 1990.

The confrontation involved one of his favorite targets, not Alex Pacheco of PETA but Shirley McGreal of the International Primate Protection League. The conflict began in Atlanta, headquarters for the U.S. Centers for Disease Control. CDC has a mission to protect the public health. It had just learned that a rare virus was being imported into the United States, via Asian monkeys. The virus had turned up in 1989 and been identified—at Fort Dietrich—as an Ebola-like virus.

As far as anyone knew, Ebola meant bad news; it meant death. In Africa, Ebola viruses are savage killers. Two outbreaks in Zaire and the Sudan had leveled villages, claimed nearly a 90 percent human mortality rate. As it turned out, the monkey virus was a monkey killer only. But CDC didn't wait for that to prove out. The agency locked down imports. Then its inspectors put on their white gloves and began dusting the shelves of every importer in the country.

In the spring of 1990, CDC sent a team down to check out the facilities of a Miami importer named Matthew Block. Through his company, Worldwide Primates, Block imported wild monkeys. One of his customers was the Tulane primate center. The white gloves were out when CDC swept through Block's facility. Inspectors listed 46 problems, among them: monkey cages stacked on top of each other, animals urinating on the heads of those below, crawling flies, waste draining from dumpsters. Block lost his import permit because of that list. In fact, almost every import operation in the country lost, at least temporarily, its permit. Although Block was able to get his license reinstated, he was not able to keep his problems private. McGreal had been tracking Block's business for years as part of a watch on primate importers. When she learned his facility had been inspected, she got a copy of the report through the federal Freedom of Information laws.

She mailed a copy to Gerone with a terse note, three lines in all: "Should [you] patronize the company, Worldwide Primates, we invite you to peruse this animal dealer's notice from the Centers for Disease Control suspending his license to import primates."

McGreal didn't expect Gerone to like the letter. She knew he didn't like her. She'd allied her organization with PETA during some of the court battles over control of the Silver Spring monkeys. Gerone had made it clear that he considered any correspondence from her purely harassment. "She's one of a group of people who'll do anything to make research look bad," he says. "She hands out numbers. She distorts facts. I used to write her back but I learned not to waste my time. She doesn't listen to what I say."

Still, stubbornly, McGreal wanted Gerone to see that unflattering CDC report. Since he bought monkeys from Block, she wanted him to consider the source. He would blow her off, she knew, but he might pay attention to CDC's concerns. What she didn't know, at the time, was that Tulane's relationship with Block was unusually tight. The center was trying to get the sooty mangabeys out of Africa for researcher Robert Gormus's leprosy project. The mangabeys, elusive and rare, were not easy to get. They weren't like macaques, in demand by the tens of thousands. There was no mangabey pipeline; it was hard to even know where to start.

Tulane had turned to Block, one of the few American dealers in African monkeys, for help. He had provided them with sources, helped with the import regulations, giving them a place to start. When Gormus eventually was forced to search in Africa himself, Block acted as banker, holding funds for Tulane, providing the scientist with a Worldwide Primates American Express card.

So when Gerone received McGreal's letter, he figured Block would appreciate knowing what she was up to. He faxed the letter off to Block. Block had his own reasons for not liking McGreal. This wasn't the first she'd made a point of publicizing his problems. At the time, she was making a crusade of linking him to the illegal smuggling of six baby orangutans, a desperately endangered species, out of Asia. He would eventually be indicted and plead guilty to felony violations of international law to protect endangered species. McGreal's efforts to publicize his role had included getting the BBC interested in the story. In 1990, the British network aired a documentary called *The Ape Trade* which strongly hinted that Block was responsible for the smuggled orangutans. Four of the six baby animals had died; they had been packed, without food or water, into rough wooden crates, labeled "birds." McGreal appeared in the program, sitting in front of her computer, blond hair brushed smoothly onto her shoulders, deploring the wild ape trade. Block appeared, too, being chased down a Miami street by a BBC cameraman.

Block received the copy of McGreal's letter in June 1990. Then he sued her. He demanded a minimum of $500,000 for her efforts to damage his business.

"She should have known better than to send it to me," Gerone says. "She should have known I would send it to him. She's made a conspiracy out of it where there wasn't one. I had no idea what he would do with the letter, I just thought he should see it. But I have to admit, I wasn't disappointed when he sued. She's constantly sticking her nose into business she doesn't understand or doesn't want to understand."

In a curious way, McGreal was more angry with Gerone than Block. She knew Block hated her; she'd been told he'd expressed a wistful desire that she'd end up packaged in a large number of tin cans. She didn't expect anything good out of him. Block was a private businessman. She knew how far apart her priorities were from his; she saw monkeys as valuable lives and she figured that he saw them as money. By contrast, she thought Gerone had a duty to be open-minded. He was a scientist supported by public money. He was supposed to be interested in free discussion. He was supposed to respect the truth. She hadn't, after all, even commented on the CDC report. She had done little more than mail it to him.

"It was the first time I realized that there were researchers who really did want to destroy us," McGreal says. "Most people would have been disgusted by the fact that these monkeys were urinating and defecating on each other. Most people would have at least wanted to know. But he, he just wanted to get me in trouble."

Block might have pursued the lawsuit to the end if hadn't been for those baby orangutans. On February 17, 1992, Miami authorities formally indicted him on charges of illegal traffic in endangered species. Three days later, he abandoned the lawsuit against McGreal, without explanation. McGreal believes that publicity surrounding the suit against her did him little good. She told everyone she knew that, when her lawyers sought documents from Worldwide Primates, Block's lawyers objected to the requests 49 times, on the grounds that the paperwork might be self-incriminating.

Block has steadily denied all her accusations and any involvement with the smuggled orangutans, which were apparently bound for the Soviet Union. Despite that, in April 1993, he was convicted and sentenced to 13 months in prison and a $30,000 fine. The case is under appeal. McGreal did her best to make sure he got a tough sentence. At one point, Miami prosecutors had agreed to let Block plea bargain to misdemeanor charges. Outraged, McGreal organized a letter-writing campaign to the court that expressed international dismay at that idea. The judge rejected the plea bargain and ordered the felony trial. With Gerone, though, McGreal has found no comparable way to slap back. She can accuse and she can storm, but there he is, down in Louisiana, running the primate center, backed by his administration. To her, he seems frustratingly untouchable.

"He says that he thinks people are so much more important than animals," McGreal says. "Well, how about me then? I'm a person. What about all the stress on me? Obviously, his circle of compassion doesn't include me or my family. No, he has to spend all his time painting me as a villain. He wants someone to hate."

Martin Stephens, of the Humane Society of the United States, calls it backlash time: "If you can believe it, even people like Taub are undergoing a renaissance," he says. "They say he did nothing wrong, that he was just set up by Pacheco. They can't even recognize their own bad apples, yet they're constantly insisting that they should regulate themselves. They believe that if they criticize their own, they're giving aid and comfort to the enemy. And we're the enemy."

The tactics of either side can seem remarkably similar. Consider the way that animal advocacy groups make scientists sound like the Dr. Strangeloves of biology, torturers of animals. They picture them as near robots in white coats, so hooked on discovery and Nobel Prizes that they can't see the animals before them. The trick in creating that image is to tell the truth, but only in part. When talking about Peter Gerone, representatives from animal groups like to whisper that he conducted biological warfare experiments on defenseless human prisoners. It sounds as if he was secretly injecting sleeping men with some fearful and deadly organism, concealing their bodies in the night. The whisperers tend not to mention that Gerone was studying the common cold, busily weighing piles of snot-soaked facial tissue. When In Defense of Animals printed a pamphlet attacking Seymour Levine at Stanford, the list of his "crimes" included "tormenting ... dogs." The description is also something of a stretch; the only dog experiment he ever did was to analyze some blood samples, shipped from Minnesota by a friend who was studying canine behavior.

The scientists return fire by calling animal activists fringe-element crazies—anti-knowledge, anti-intellectual, and unable to place any value on human life. Like their opponents, scientists have learned that half-a-truth can be better than a whole one. The most famous example is a statement by Ingrid Newkirk of PETA. In 1986, she was discussing animal suffering with a Washington reporter: She put it like this: "When it comes to feelings like pain, hunger, and thirst, a rat is a pig is a dog is a boy." It's only the last 11 words of that sentence that ever get repeated by animal researchers: "a rat is a pig is a dog is a boy."

The translation is simple: See, they're nuts. They think a nasty disease-ridden rodent and a farmyard animal are equal to a child. Pete Gerone, in fact, keeps a list of Newkirk's quotes in his computer to print out when he gives speeches. His favorite is the abbreviated rat-boy quote, which appears twice in his notes. His second favorite Newkirk quote is "Humans have grown like a cancer. We're the biggest blight on the face of the Earth." He figures that one really offends people. Gerone chortles over Newkirk's way with words. "I think it's obvious that Pacheco is the brains of PETA," he says. "Because she keeps saying things that are actually better for our cause than hers."

PETA keeps a file on Gerone as well, transcriptions of talk shows and speeches he has done; copies of his correspondence with PETA members and governmental agencies. He quotes Newkirk; they mail out copies of his letters—each hoping to make the other look bad.

One letter, sent by Gerone to the U.S. Department of Agriculture, and carefully filed by PETA, vividly illustrates the tension. In March 1989, a USDA inspector had toured the Tulane primate facility and issued a list of minor complaints; rotten fence posts in outside runs, chipped and worn floors inside, a mouse hole in the guinea pig room, a monkey crammed into a cage too small for him. Gerone fired off a letter to the Washington, D.C. administrator for all inspections, furiously pointing out that activists could get a copy of the report. Shirley McGreal has a trailer parked outside her office, crammed to its metal roof with papers gathered through Freedom of

Information Act requests. She could use these complaints to take cheap-shots at his facility.

"The point I am making is that USDA, without intending to do so, is playing into the hands of the animal rights/anti-visectionists whose stated goal is to abolish animal research. Such a report will make the Ingrid Newkirks of PETA and the Shirley McGreals of the International Primate Protection League very happy!" Gerone noted that scientists tended to regard USDA as friendly to research. He added that he had given a speech in April, and "As I said in my speech, ... if these are our friends, we don't need any enemies."

Why couldn't USDA just sit down and discuss the problems over lunch? After all, Gerone argued, none of the complaints had anything to do with keeping healthy monkeys: "Our monkeys are getting better medical treatment than probably two-thirds of the world's population." He ended with a warning: "I am convinced that if you are trying to placate the animal activists by nit-picking inspections, you are engaged in an exercise in futility and you will only serve to do us irreparable harm. You will only satisfy them if you decided to ban animal research."

There are times when animal advocates, in their indignation, sound almost like researchers of a decade ago; a sort of how-could-they do-this-to-us tone permeating their conversation. At In Defense of Animals, Suzanne Roy actually assembled a memo called "Dirty Tricks by the Research Establishment." One of her primary complaints was that researchers keep trying to make animal activists look trigger-happy, accusing them of crimes they have not committed. For instance, in February 1990, the dean of the veterinary school at the University of Tennessee was shot to death in a still-unsolved killing. The sheriff's department reported that biomedical researchers promptly called up to ask if an animal activist had gunned the man down, to suggest that was worth investigating. Among those who called was Frederick King, director of the Yerkes primate center, who says he was just expressing concern and curiosity. After all, he had received numerous death threats from animal advocates. The calls turned into a rumor that some animal-nut was responsible for the killing. The AMA then reported the rumor as further evidence that animal rights activists were terrorists.

"There are science advocacy groups out there that do not seem to have a high regard for the truth," says Stephens, at the Humane Society of the United States. "They misquote, they make up quotes. They're really bending over backwards to polarize the situation."

Officially, his organization tries to be neutral to the point of blandness on animal research, to maintain a negotiating position. The formal HSUS statement on the issue "recognizes that some scientific research and testing on animals has benefited humans and animals and that such research continues to rely on animal subjects to a large extent."

But again, like animal activists, scientists have come to have long and unforgiving memories. Back in the mid-1980s, the Humane Society was taking a tougher stand. One HSUS officer stated publicly that torture was a founding principle of experimental psychology, offending many scientists. He suggested the whole field could be disbanded without harming science. These days, Stephens can talk neutrality all he wants, but what scientists remember is that, less than a decade ago, his group sounded as if it was advocating the abolishment of a major field of research.

"We're constantly having to defend who we are," Stephens says. He finds himself, some days, satisfying no one. Oh sure, say the scientists, if you're so moderate, let's hear you come out and make some public statements in support of animal research. "He [Stephens] should sound defensive," Gerone says. "Because we're going right to the crux of the matter. Let's see them come out publicly for research. I've written them letters to that effect and they don't even write me back. I'll believe they're moderate when I see that in writing. And I'll never see it."

He's right. The membership of the society wouldn't stand for it. In the spring of 1993, Stephens was asked by a national magazine for his opinion of groups like ALF. He condemned acts of violence. He also tried to explain to the reporter that scientists may also be troubled by animals suffering in research, that a painful experiment is often viewed as a necessary evil. The reporter quoted Stephens himself as believing that animal research was a "necessary evil." Angry mail came pouring into the society's headquarters—how dare he speak out in support of animal experiments, how dare he even try to defend researchers?

"I don't feel that it's the position of an animal protection organization to endorse animal research," Stephens says. "But the hardliners push for it all the time. Especially the older guys, like Gerone. They grew up doing science at a time when no one questioned what they did. They haven't gotten accustomed to the change. And the buzzword these days is get out on the stump and talk up animal research."

At first, it was just a few angry scientists, Gerone, Frederick King at Yerkes, Adrian Morrison from Pennsylvania. Then others came in, Stuart Zola-Morgan, from UC-San Diego, with his explorations of the brain and his determination to defend them. Rick Van Sluyters, a neuroscientist from UC-Berkeley, who studies vision in cats, was another. He had been informed by local police that they had uncovered a plot to bomb his home.

And then the resistance, as it were, began to organize. With neuroscientists so obviously under attack, the national Society for Neuroscience formed an animal research committee. It has been headed by Morrison, Van Sluyters, and now Zola-Morgan.

In 1989, copies of an American Medical Association document, titled an "Animal Research Action Plan," were leaked to animal advocacy groups. In her organization's monthly newsletter, Shirley McGreal described the AMA White Paper as the "AMA's plan to destroy the animal protection movement." She fired off a letter to the American Civil Liberties Union, stating the document seemed to run counter to the Constitution. Further, she insisted, it had the potential "for creation of an atmosphere in which people will be fearful of expressing their opinions on behalf of animals."

The AMA plan was among the first to suggest labeling animal activists as "anti-science and anti-progress." It advocated the kind of emotional appeal that Pete Gerone likes to use: Your child or a drowning rat? It suggested that scientists meet with the media, to warn of the real threat of animal activities. It advocated heavy pro-science advertising. The plan also explored the possibility of legal challenges to the nonprofit status of animal rights groups, building a private data base on animal rights activities and intensive lobbying on animal research legislation.

That torch had already been taken up, however, by the National Association for Biomedical Research and its president Frankie Trull, a woman considered both politically savvy and articulate. The national association started small in 1979. In the mid-1980s, it consolidated with the National Society for Medical Research—an older group, one which had given Christine Stevens problems. The new group began to build a power-base. More than 350 institutions— universities, medical schools, vet schools, teaching hospitals, academic societies—are members. NABR is nonprofit; its annual income is about $800,000. The group gives the lie to the idea that scientists are politically naive. It is headquartered in Washington, D.C. Its annual reports are dotted with photos of legislators and sprinkled with lists of lobbying successes: federal laws that raise the penalty for attacking a research institution; defeat of a bill that would restrict the use of animals to test product safety.

NABR, through its affiliated Foundation for Biomedical Research, has trained scientists in media awareness, showing them, among other things, how to find loopholes in Freedom of Information requirements. It has held workshops on libel law. Trull has also trained researchers to be prolific letter writers: "In the past five or six years, researchers have been great about Capitol Hill," she says. "The letters come by bucket-loads. They've learned to be more politically active. I think politics used to intimidate them. It doesn't anymore."

The foundation has also countered animal activist photos of burned dogs and cats, wires protruding from their heads with a series of heart-tugging posters. The posters boil it down to Pete Gerone's rat versus baby argument. It's not a subtle argument, and neither are the posters. A typical one is a pretty, blond little girl (the foundation does not use elderly, fat people as models) surrounded by stuffed animals. "It's the animals you don't see that count," reads the text. The implication, not hard to get, is that the child would be ill, if not dead, without medical use of animals.

In a curious way, Trull can sound a lot like Stephens; she works hard to present NABR as a moderate organization. "We don't tell our members how to think," she says. "But just because science did something, doesn't mean it should be defended. The only way to maintain integrity is to get the facts and wave them." The association's official credo supports humane care of animals, keeping the numbers used to a minimum, and that "alternatives to the use of live animals should be developed and employed wherever feasible." Her weekly newsletter is informative rather than shrill. When President Clinton picked physicist John Gibbons as a science advisor, biomedical researchers were shocked to learn that Gibbon's wife had once been a member of PETA. People were actually calling each other long distance to disclose that Gibbons had an old PETA bumper sticker on the family car. Rumors started flying immediately. There were fears that the administration was about to dismantle research, require extreme animal protection. Trull calmly reported that the Gibbons had let that membership lapse—deciding PETA was too extreme. She added, however, that the family now belongs to McGreal's IPPL, which had been known to ally itself with Newkirk and Pacheco.

If scientists doubt the intentions of the Humane Society, activists are equally disbelieving of Trull's insistence that NABR is even remotely moderate. "It's become an awful group," says Christine Stevens, of the Animal Welfare Institute. "They thrive on 'us' against 'them.' " Stevens blames NABR for helping to destroy decent regulations in the 1985 Animal Welfare Act. When the USDA announced that it would toughen earlier regulations, NABR filed an 118-page legal brief in opposition to the changes. After the agriculture agency backed down, issuing rules far more to the liking of the science community, Trull wrote the agriculture department a congratulatory letter, offering to provide any information needed and adding that "the association looks forward to continued cooperation with you."

When the regulations were overturned in court, however, USDA did not rush to appeal. There were rumors in animal research laboratories that the Clinton administration was pro-animal rights, would let tough regulations stand. In that case, NABR made it clear that the group would not abide by USDA's decision. It successfully petitioned for standing to sue on its own. Further, the group lobbied ceaselessly for USDA to appeal anyway. The agency finally agreed to fight in late 1993. Stevens blames NABR largely for that: "They just pushed and nagged and pushed until the government had to do something." In announcing its 1993 annual meeting, NABR warned that the country is in the midst of political change and that "while change and reform may be necessary in some areas of science and research, the progression should not be dictated by the animal rights movement through political, legal, and social manipulation."

Such success has led to a rising tide of baby NABRs. There are 17 state organizations, all pushing for easier use of animals in research. The groups grew out of places like California, Michigan, and Massachusetts, where scientists have been badly rattled by aggressive animal activism. There are a host of full-size, national competitors too.

The Connecticut-based company, U.S. Surgical Corporation spent almost $1 million in 1991 to charter a new group, Americans for Medical Progress, that billed itself as an educational foundation. The company's chairman, Leon Hirsch, has been an outspoken crusader against animal rights and an unyielding defender of the need for animals in research. His $800 million-a-year company makes surgical supplies, such as staples, that have been routinely tested on dogs. The company ran its own pro-research publicity campaign for several years, running ads even less subtle than NABR's. "Here's the new Guinea Pig" declared the caption under a picture of yet another pretty blond child. The tone of Americans for Medical Progress has been clear from the start. The group recruited such outspoken primate researchers as the Yerkes Center's King and Douglas Bowden, head of the Washington Regional Primate Research Center in Seattle, to serve on its board.

In January 1993, the group sent out a letter, from Frederick King, recruiting new members. The letter was not calm. "It is clear to me that if the research community does not stand up for itself, research will slowly die," King wrote. "It is essential that the research community defend itself against the so-called 'animal rights' movement." Like Gerone, AMP has also learned some valuable lessons from animal advocacy groups; that you get strong support by stirring up strong feelings. In 1992, the chairman of its "Jewish Task Force" mailed out a "Dear Friend" letter, accusing the animal rights movement of vicious anti-semitism, describing the movement as opposed to biomedical research and Jewish kosher animal slaughter. "As a concerned Jew, whether or not you keep kosher, you have an extra special reason to put these organizations out of business. Because they are interested in putting Jewish values and practices out of business."

The message gets across. When Matthew Block was facing sentencing on the orangutan smuggling charges in early 1993, a furious letter of support appeared in the New York-based publication *The Jewish Press*, saying: "An outrageous miscarriage of justice is being perpetuated against a wonderful Orthodox Jewish husband and father of two small children by Jew-hating animal rights extremists."

The success of such organizations has inspired still more. Many consider the most hard-line to be Putting People First, a 40,000 member lobbying group headquartered in Montana. The group combines an unswerving support of animal research with a dislike of "environmental extremism." Putting People First chairman Kathleen Marqueth writes a column in *Fur Age Weekly*. Her stance against the overprotection of whales and in support of the rights of whalers has earned the group a spot in Greenpeace's list of anti-environmental organizations. In contrast to Trull's rather low-key account of Clinton's science advisor John Gibbons, Marqueth's was downright hostile, complaining that PETA now had a defacto office in the White House, warning that "Clinton may rue the day he appointed Gibbons." Trull herself describes the group as "at the opposite end of spectrum from PETA."

Bureaucracies and organizations have one major advantage over individual scientists. People, on their own, burn out. That Pete Gerone, for instance, has not flagged in his crusading zeal has a lot to do with his university's unflagging support. He knows it. He says scientists come up to him, speaking enviously of the backing he gets from Tulane president Eamon Kelly. Kelly has stood by Gerone through lawsuits, protests, and hate mail. That isn't always true. Administrators, in general, loathe controversy. The animal rights/ animal research issue is loaded. Sometimes its hard to get officials to even support their own faculty, much less take on the broader cause. When the ALF fire-bombed the offices of a mink researcher at Michigan State University, the university's president never met with the staff there, to offer comfort.

Stuart Zola-Morgan, at UC-San Diego, tells of a graduate student at another state university campus who wrote a letter to the local newspaper, criticizing animal rightists. They sued him. The university lawyers told him to issue a public apology. He hired his lawyer and got the lawsuit dropped. "But it sent a clear message to us all that we were on our own," Zola-Morgan said, "that the institution would only back us so far."

And for all his determination, Frederick King at Yerkes offers one of the best examples of wearing out in the service of defending animal research.

Starting in the early 1980s, King became one of the most outspoken of the critics of the animal rights movement. He gave speech after speech: "Extremists in the animal rights movement probably will never accept justification for research or assurances of humane treatment. But, fortunately, there are many who, while deeply and appropriately concerned for the compassionate treatment of animals, recognize that human welfare is and should be our primary concern." When journalists called, he was always ready to sound off. When Frankie Trull at NABR asked him to stump for animal research, he did it; he recalls doing nine newspaper, television, and radio interviews in Detroit in two days, some of them more than two hours long. He constantly pushed other scientists to step forward, arguing that science would have far greater public support if researchers would just get out there and tell the public what they were doing, what the tax dollars were paying for, why it was important. "The public must be told that it is scientists who occupy the middle ground," he argued, not animal rightists, who seek to end research.

"I became one of the first spokesman, partly because one of the first demonstrations was held at Yerkes. Three hundred people marched here," he recalls. "Most scientists then thought, 'This will never happen to me, it will go away.' I didn't believe that. I saw that the animal rights organization had tremendous potential for developing power and effectiveness if it had good leadership. I thought we had to become proactive immediately."

"People in the animal movement call me as an extremist, I realize that. I don't think I am. But the issue has become increasingly polarized. My experiences have taught me that the leaders of the opposite side, and people like me, are not going to be able to talk effectively together and accept each others' basic philosophy."

For example, King says, the question of whether animals have rights: "I don't believe that rights fall like manna from heaven. Rights are not magical or absolute. I know our Constitution talks about inalienable rights. But I haven't seen any evidence of these. Rights are given by one group to another. When a group cannot accept responsibility for the laws governing the community, then rights are denied. We deny rights to minors and children, to people with certain kinds of illnesses, certain rights are denied to people who violate the law.

"And animals, to the best of my knowledge, are unable to comprehend the rules of society. If animal rightists had their way, we'd open the doors to zoos, let dogs and cats out into the street. Does that make sense? It's not animal welfare and it's not wise. But if you believe that animals magically have the same rights as humans, but that they simply can't speak for themselves, then of course you're going to be opposed to animal research. And that gap, I think, is almost unbridgeable."

He is determined in that belief—and he is weary. King plans to retire as director of Yerkes in the fall of 1994, although he will remain on as a staff scientist. In the past few years, he has consciously pulled himself back, tried to be less of a public combatant in the fray over animal research. His own visibility has made Yerkes a constant target of advocacy groups. It seems that everyone has taken their shot, even at the most benign projects.

When the center took on a study of the sexual attributes of gibbons, trying to understand why the tree-swinging apes are monogamous, as opposed to other apes, like chimpanzees, McGreal publicly ridiculed it as "Project Penis." Atlanta-based Friends of Animals charged the center with mistreating its chimpanzees, claiming that keepers disciplined the apes by firing BB guns at them. King was forced to hold a press conference to deny the charges, saying that he had outlawed such tactics 14 years ago when he became center director. In April 1993, Last Chance for Animals staged a hunger strike in front the center, protesting its abuse of primates.

It gets old.

"For a long time, there were very few of us out there. And I got kind of tired. I thought it was time for others to pick up the banner. I think some of the new people are more effective than me now. In the early days, it was more banging heads than it is now. I think perhaps my style has become a bit archaic. Maybe I started to get a bit too combative. Sometimes that can do more harm than good. We really didn't understand that then. I think we got targeted more than we would have otherwise."

Emory University, which runs Yerkes, has come to regard him as a lightning rod, he says. The administrators there don't like it; they worry that controversy is bad for the school's public image. King suspects that the animal advocacy groups have tried to exploit that division, turn Emory from Yerkes, isolate him from support. He suspects, too, that the next director of the primate center will be less of a brawler.

That's not to say he believes the war is over. More, he thinks, it is shifting ground, that animal advocates are targeting the future, the next generation of voters and scientists. He's come to believe that public speaking may be almost a waste of time, that he's either addressing the already converted or people who are committed to the other side of the philosophical gap. The new fighting ground will be in the schools.

You can start wondering if there's anyone in this issue who isn't conducting an education campaign. The battle to win over school kids has become so intense that, if teachers chose (and if they could get approval), they could spend their weeks teaching nothing but animal research issues. PETA is doubling its school education budget to more than $1 million a year; In Defense of Animals now has a full-time staffer who goes from school to school, lecturing and giving away buttons and brochures supporting animal alternatives; the Humane Society has established a center, staffed by 14 people, to develop educational materials; last year, the 400,000-member American Society for the Prevention of Cruelty to Animals also set up an education center.

NABR has films and posters designed for school-age children.

Putting People First, which has about a $225,000 annual budget, has announced that it will develop a kindergarten-to-twelfth grade curriculum covering the history of the relationship of human beings, animals, and biomedical research. Even the state groups are targeting children; many groups, such as Michigan and California, now hold pro-animal research essay contests for school children, with cash prizes reaching $500. South Carolina's Association for Biomedical Research passes out a coloring book, The Lucky Puppy, about two children whose dog is saved, thanks to medicines developed through research on rats. The book is cheerfully upbeat about the joys of animal research. At its conclusion, one of the children happily declares "I want to be a research scientist. Then I can help animals and people!"

In its simplicity, it calls up the image of Pete Gerone telling peg-leg pig jokes at the dinner table, arguing that surely, surely the farmer comes before the pig. Yet, it is in this arena that Gerone is less confident. He is relying on his own generation. He even thinks the baby boom generation, largely, believes in the cause of animal research. He is far less sure about the next generations. The audience is different: less trusting of science, more skeptical of government intentions, more doubtful of human superiority. He worries about the people who don't laugh at his joke. He's not sure how you win over children who have never seen a farm, only known pets. Or how you attract students who think of science as either baffling or dull, a troubling pattern in American schools. "We're losing the fight in education," he says. "The animal rights people are better at it than we are. I worry about tomorrow's minds. I worry that they won't come to us."

Arthur, a young chimpanzee at the Yerkes Regional Primate Research Center, part of a study on how best to rear youngsters if they cannot be raised by their mothers. *(Photo by Cathy Yarborough, Yerkes Regional Primate Research Center)*

Duane Rumbaugh with young chimpanzees at his Language Research Center. *(Language Research Center, Georgia State University)*

Rhesus macaques gather during the day in one of the open enclosures at the Yerkes field station. *(Photo by Cathy Yarborough, Yerkes Regional Primate Research Center)*

Young macaques in a Harlow study of curiosity and motivation in monkeys. *(Courtesy of Harlow Primate Laboratory)*

A young rhesus macaque plays in an incubator at the Yerkes Regional Primate Research Center. The monkey was part of a study to determine the postnatal effects of cocaine exposure. *(Photo by Frank Kiernan, Yerkes Regional Primate Research Center)*

Bottle feeding of an infant baboon at the Southwest Foundation for Biomedical Research. *(Southwest Foundation for Biomedical Research)*

Baboons in the big breeding colony at the Southwest Foundation for Biomedical Research, San Antonio, live in an open corral. *(Southwest Foundation for Biomedical Research)*

Squirrel monkeys, tagged for identification, at the California Regional Primate Research Center. *(Photo by Patrick Barreto)*

Two crab-eating macaques huddle together in paired housing at the nursery of the University of California, Davis, Primate Center.
(Copyright ;© Lois Bernstein/

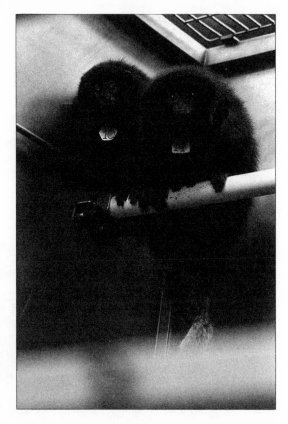

Two titi monkeys sit, tails entwined, at the California Regional Primate Research Center.
(Photo by Patrick Barreto)

A sooty mangabey at the
Abidajan Zoo, Ivory Coast, 1991.
(Courtesy of Robert Gormus)

Robert J. Gormus with an infant sooty mangabey named Na-Na in Accra, Ghana.
The monkey is now, with Gormus, at the Tulane Regional Primate Research

SEVEN

Hear No Evil

"LET ME TELL you a funny story," says Susan Lederer, her voice bright and amused over the telephone.

Lederer, at Pennsylvania State University, is a science historian with a passionate interest in the early twentieth century. She's particularly fascinated by the tug-of-war between researchers and animal activists, up to a point. She says frankly that—for now—her scrutiny stops about the time Hitler began making bloody inroads in Europe. She's writing a book on just that theme, with the working title: "The Scientist at the Bedside: Medical Experimentation and Anti-Visection Before World War II."

In 1992, the University of Alabama in Birmingham invited her to come down and talk about her work. A few weeks before the lecture, she received a mildly panicked phone call from administrators at the southern university. What exactly was her specialty again? "Animal research," Lederer replied. An agitating buzzing filled the line: "Oh, no, no. Too controversial. What else do you study?" Well, Lederer said, she also wrote about medical experiments on humans. Great, said the Alabama official, not controversial, no problem at all.

Lederer bursts out laughing as she tells the story. "So I went down and I gave a public lecture on human experiments and closed-door one on animal experimentation. Doesn't that seem strange?" She suspects, however, that the University of Alabama is unusually sensitive on the topic; it is, after all, the academic home—or, if you prefer, sanctuary—of Edward Taub, of Silver Spring monkey fame. Anyone connected with Taub has seen more than a share of controversy, not to say name-calling. "Still, I've never had the experience of having a lecture rescinded before," she says. "No one ever said why. I didn't even know Taub was there until I read it, later, in *The New Yorker*. No one in Alabama breathed a word about it."

There's an obvious irony in this: as a historian, Lederer looks at how animal researchers keep the grisly side of their work from the public. If you think of scientific results like a photograph, then Lederer is interested in the airbrushing, the smoothing out of details that might alarm the viewer. She focuses on word choice, the careful use of jargon to sometimes obscure meaning. And although her interests, for now, are strictly pre-World War II, the issue remains very real today. In the war for minds—the one Peter Gerone worries about losing—the major weapon is information. The winner is the one who controls it.

Scientists, Lederer points out, learned that lesson early. She can document it with an extraordinary resource from the 1930s, the internal files of the *Journal of Experimental Medicine*. The journal's editor, Francis Peyton Rous, was ruthless in his insistence that scientists protect themselves from attack whenever possible.

Under his direction, scientists could not say animals were starving. Instead they were "fasting." Experimental animals didn't "bleed"; they "hemorrhaged." They didn't receive "poison"; they were given an intoxicant. Anything too graphic, by way of description, was taken out. Lederer cites the case of a 1934 article in which the original draft mentioned the "almost continuous moaning" of ferrets being used, "whining as though the animal was in pain." The journal deleted the sentence from the published version. In a 1935 article on blood pressure readings from anesthetized cats, Rous had the text changed from "the brain was sliced off at various levels" to "the brain was removed." In a 1938 letter to a researcher, explaining why "pronounced" had been substituted for "severe" in the description of an injury, Rous wrote: "severe, as applied to a condition of animals, always requires justification, and while your experiments can readily be justified, there is no reason for provoking a need to do it."

Lederer has a historian's affection for Rous, who was a notorious pack-rat: "He was remarkable; he saved everything." She sees him also as representing a twentieth-century trend, using language as cover. In the nineteenth century, scientists were more forthright. "They were really out there, back then. They'd write that they had to kick a dog in the stomach until it was whining in pain before they could do the procedure. They had much less sense of how it would play." Perhaps they didn't care either; in their time, scientists were less intent on keeping up a good public image. But continued pressure from animal advocates changed that, she said. Rous's strategy is more than evident today. One modern example is the way biomedical researchers talk about animals' deaths. They may "sacrifice" animals; they may "euthanize" them. But they do not "kill" them. And writers who use the word "kill," including this author, are criticized by research advocates as inflammatory.

There's a short-term gain in this, Lederer believes, but a long-term error. Certainly, careful language can keep an experiment from drawing attention and attack. Harry Harlow, with his "pits of despair," is an intimidating example of what happens to a researcher who errs too far in the opposite direction. Still, Lederer thinks that such cocooning of truth with euphemism is risky. It insulates the public from what science really is. Then, when presented with the sometimes grim realities of experiments—gruesome pictures furnished by animal activists—people are shocked and dismayed. "If scientists are willing to pay the price—that the public doesn't really understand science—that's one thing," she says. "But I don't think they are. And I do think that's what's at stake."

In this era of belligerent animal advocacy, that question has grown far beyond a small, academic debate to one of the more serious in animal research: What is the truth? Who gets to tell it?

Animal activists have done their best to claim that role—by making as much noise as they can, telling their side at top volume. Slowly, researchers have realized that Rous's approach isn't enough. They have learned to shout back and they have learned, on occasion, to stifle the opposing view. They have also learned to decide what the public should hear. Organized and angry, scientists have forced changes in network television shows and popular encyclopedias. They have barred people with unpopular opinions from getting published. They have threatened dissenters' jobs. The desire on both sides to win over the public has led to bitter lawsuits, to longstanding grudges. To follow this ever-widening circle of influence, a good starting point is perhaps the obvious one: A library, paid for by taxpayers, intended as a place where information is available to all seekers.

This particular library is a special one. It is the book-lined sanctuary of Larry Jacobsen, chief librarian at the Wisconsin Regional Primate Research Center in Madison. The center sits squarely across the street from the Harlow laboratory, the two of them presenting a study in contrast. The old lab is aging, industrial in character; the center is upright, modern, painted a spring-like pale peach.

The library, on the center's second floor, is sunlit, with sofas set against windows, antique cabinets holding first-edition books, tapestries depicting monkeys dancing hanging on the walls. Jacobsen is fair-haired, blue-eyed, and gentle in his speech. Away from the center, he lives in a beautiful old Mission-style house, a polished wood setting for his collection of Mission furniture. He gathers friends together before the carved fireplace and serves brandies on a silver tray. It's not just that the home is lovely, but that it's lovingly put together. That's true of the library, too. There are 5,000 slides in the primate library, 500 videotapes, almost 18,000 books and journals, audiotapes, films. Stephen Suomi, at NIH, calls it the best primate library in the world.

But it isn't about numbers and it isn't about storing documents. It's about sharing them. Jacobsen loves the sharing of information, of ideas. He's headed the library since 1973. To him, it serves as a transfer station; the information that comes in should go back out. The staff here put together the first *International Directory of Primatology*. It offers the most detailed listing anywhere of who has captive monkeys and why. The center offered it for a bare-bones cheap $10 and sold out. As primate researchers began more and more to share information by computer network, through electronic mail, Jacobsen put together a communication network for primatologists, Primate Talk.

The network is barely two years old. It has almost 400 members in 18 countries. Primatologists share medical information. They share data. They once launched into a nearly two-week-long discussion of whether humans and apes could interbreed. (One scientist jokingly suggested that the rare albino gorilla at the Barcelona Zoo was an example of such an experiment. That prompted a snappy denial from the zoo's director, also a Primate Talk subscriber.) Scientists are forthright in their discussions on the network. In a lengthy debate over why so many rhesus macaques are used, one Mississippi researcher announced that he used them because, although they are vicious animals, they were not as vicious as apes. People have complained that Primate Talk can get really silly. It once featured a lengthy commentary on Barbie dolls. To Jacobsen, the point is that people talk to each other.

"It's a leveling ground," Jacobsen says. "Graduate students get to talk to top scientists in the field. There are people who are the only primatologists at their institution. They've told me that they've suddenly felt connected, that the first thing they do in the morning is get a cup of coffee and look for the new messages on Primate Talk."

That's the ideal. He's known for years that he wasn't going to always possess the ideal, not as director of a library in a university primate center, funded by the federal government. He dates his own awareness to 1975, when a pair of young men ambled into the library. At first, Jacobsen was just curious about who they were. And then he was alarmed. At that time, the center stored all its necropsy reports in the library, every detailed analysis of every animal that died. The newcomers went straight to the reports and began photocopying them: click, copy of dead animal record; click, copy of dead animal record. Jacobsen listened uneasily to the steady whir and shuffle of machine and moving paper.

"It was just at the beginning of the animal rights movement," he recalls. "I just suspected. I'd never seen these people before. And these were very technical documents. I didn't know what to do. I felt really torn. I felt at least an obligation to tell the director and let him make a decision. It's funny, I couldn't make a decision myself." Jacobsen walked over to the office of the center's head, then an easygoing, friendly psychologist named Robert Goy. He told Goy about the necropsy reports. He remembers watching Goy's face turn bright red. The director walked back with him to the library. Goy happened to be a man who believed, resolutely, that fairness was imperative. He did not rip the photocopies away. He told the men they could make one more copy and leave with what they had. That week, the center staff voted to remove the necropsy reports from the library, where the public had access to them.

"I really think librarians have a responsibility to make both sides of the issue known," Jacobsen says. "But for us there's this underlying issue—we're a primate research center, funded by NIH. And NIH isn't interested in funding a facility that's going to help destroy its programs." The Wisconsin animal activists did, in fact, get copies of all the necropsy reports through public record laws. They then staged a mock funeral outside the primate center, setting a black coffin at its doors, dropping a card into it for each dead monkey, reading off the monkey's lab number and it's cause of death.

"They only used one word from each necropsy report," Jacobsen said. "It would be monkey X-68, 'asphyxiation.' That was all. They showed pictures of dead monkeys, too. Some of them weren't even ours." He remembers being surprised at how irrational the attack seemed, realizing that information shared with these activists would not be used in ways he considered to be responsible. "And in this library, which I want to be democratic, equal access to everyone, I find myself in the difficult position of trying to give equal help to people I perceive to be good guys and those I perceive to be bad guys."

He doesn't always get to choose either. Sometimes Jacobsen doesn't have access. There are things he would like to put in his collection but can't get. Every researcher he ever asked for a photograph of a self-abusing animal—an isolated monkey that has attacked itself by biting, tearing out hair, picking at skin—has refused. The library has had similar difficulty obtaining videotapes of some surgical procedures, even when the individual scientist is proud of developing a new technique. The scientist will say yes; the boss will say no, too worried about animal rights abuses of the information. Jacobsen concludes unhappily that the effect has been to "launder the collection a bit."

When he was putting together the international primatology directory, he decided to include Shirley McGreal. He knew she was controversial, but after all, she did house almost 20 gibbons and give small grants for conservation research. Officials at the Yerkes primate center promptly announced that they were pulling out of the directory. They didn't want to legitimize an animal activist. McGreal has been banned from other information directories. The Washington Regional Primate Research Center has told her directly that nothing she does, not even her conservation reports, will appear in its listing, a clearinghouse of written reports on primates.

In the case of his own directory, Jacobsen argued Yerkes back in, pointing out that McGreal was clearly identified and that she was a single inclusion, that he was hardly allowing the directory to become an animal rights guide. The point mattered to him. "I'd like to see, for better or worse, the whole picture represented. If you're denied access to certain things, it means neither side gets to see them. In a country that allows free speech, after all, if someone misuses information, you have a right to try to correct what they said. I'm not saying I'm without prejudice, of course. One tries, but that's not how life is."

American science—barring patents, classified information, spreading commercialization—is born and bred in openness. When a researcher finishes an experiment, good or bad, the first rule of business is to try to publish it. Out it goes, detail by excruciating detail, allowing other scientists to check it out, expand on it, build on it. The tradition of sharing is so firmly established that when animal researchers back away from it, shelter their work, it can seem to go against the very grain of the profession. It emphasizes how afraid they are and how mistrustful. There is a sense within the research community that the animal advocates are out there, waiting to get us. It's been an almost instinctive response—they're out there, let's hide. As a result, one of the most troubling by-products of militant animal activism is the driving of science into the shadows. It pulls it not only away from critics but from the public that pays for it.

If you look carefully at what happened with the 1985 Animal Welfare Act, one of the most revealing changes came in an obscure, legalese passage about public records. The change was about allowing once-public records to become private. On this one, you have to give scientists credit. That paragraph means that they were effective lobbyists. They were able to get written into law a very neat end-run of the Freedom of Information Act.

The question of more privacy for researchers arises from another significant requirement of the 1985 law. It says research institutions must try to make their captive monkeys as content as possible, given that they are caged in a laboratory and subject to experiments. If they are part of the breeding colony or scheduled for an AIDS experiment, no difference. Whatever the monkey's status, the law orders scientists to take good care of them, beyond food and water, to provide for the "psychological well-being of primates."

The research community has been struggling with that requirement since it first appeared. As a society, we've yet to achieve psychological well-being of humans, much less other species on the planet. Do you hold out for happiness or settle for tranquility? And how do you know what you've achieved, what marks a psychologically healthy animal? The new USDA rules require each institution to work out a plan. Scientists must explain in writing their goals and how they mean to achieve them. Do they have toys? Fresh fruit? Company? Since monkeys are highly social, if a lab means to isolate them—perhaps due to an experiment with a highly infectious disease—that must be carefully noted too.

Normally, those explanations would be public property, accessible through the Freedom of Information Act. Taxpayers, who pay for care of the monkeys, could see them if they wished. So could animal advocacy groups. But researchers did not want to share that information, to provide their enemies records of how they might blunder around, trying to satisfy the mental health needs of monkeys.

So the sidestep around the information act works like this: If the primate plans were, as usual, turned over to USDA inspectors, they would be public record. But, if they were stored at the individual institutions, they would not. They would be that institution's property. If you were really determined about it, and the institution was, say, a public university, you could try to get access to them under state public record laws. If the institution was private—Stanford, Harvard—you were out of luck. And that's exactly the way the regulations work: on-site storage of the plans, available to USDA inspectors but not to prying activists. Scientists made no secret of their intention to avoid the law. One federal administrator said the research community simply wanted to avoid "unreasonable criticism by uninformed people."

Christine Stevens, who had worked ceaselessly to get the new law passed, suddenly found herself shut out of learning what it had accomplished. She was absolutely furious. The information dodge became part of the lawsuit that she—along with Roger Fouts and other plaintiffs—brought against USDA, seeking to overturn the new regulations. It was the one challenge they lost immediately. U.S. District Court Judge Charles Richey wrote that, while he wasn't happy with the sidestep, it was permissible under the law. The Freedom of Information Act, he noted, does not require agencies to operate in a way that "maximizes" the amount of information available to the public. "Therefore, in this respect, there is nothing this Court can do about this, however much it might be inclined to do so."

In fact, researchers have learned that they can exercise considerable influence over what the public sees and hears. Their interest is not limited to government documents either. They've learned that other sources—such as commercial television—can be far more influential.

In the fall of 1991, NBC planned to devote an episode of its then-popular series, *Quantum Leap*, to the issue of primate research. The premise of the series was that physicist Sam Beckett had, through a breakthrough theory, combined with some mysterious supernatural force, become a time-traveler, his consciousness ricocheting through recent decades, spending each episode in a different body. He always dropped into the body of some person—man, woman, or child—who was undergoing crisis, his mission being to foresee and avert the disaster threatening that life or one nearby. For the primate episode,

Beckett would drop into the body of a chimpanzee. It was a research chimpanzee, scheduled to go through a skull-cracking, crash impact test. Could he save it?

When a fall preview issue of *TV Guide* mentioned the episode, hinting that it would be pro-chimpanzee, researchers were appalled. In their minds, it would be Dr. Animal Butcher all over again. They had no intention of rolling with it. Frankie Trull sent out a hasty memorandum. Re: "Letters Needed to *Quantum Leap.*" She requested defenses of head-injury work; she provided the names and addresses of *Quantum Leap's* executive producer, the co-executive producer, the president of NBC, and the actors who starred in the program. Her own letter to the network insisted: "It is imperative that the public, your audience, continue to support humane and responsible animal research." The mail came from around the country; scientists accusing the network of airing propaganda for fanatics. Executive producer Don Bellisario complained to the *Los Angeles Times* of the heavy pressure. In the same article, Yerkes researcher Kenneth Gould said he would not object to a show that told both sides, but "Given the choice of an animal rights advertisement for an hour, I'd rather see the whole thing scrapped."

There were researchers who disagreed, sent in strong letters of support, among them Roger Fouts and Jan Moor-Jankowski, director of the Laboratory for Experimental Medicine and Surgery in Primates in New York. Moor-Jankowski wrote an impassioned letter to Bellisario in favor of free speech. He said that even though his own laboratory uses chimpanzees in AIDS research, and even though he might expect some negative fallout from the episode, he still thought that science was best served by open discussion. In a clear illustration of how divisive the issue had become, he also took a slap at NABR.

Trull had informed NBC that crash tests were done on anesthetized animals. There had been, however, exceptions to that. In 1973, NIH scientists had driven pistons into the heads of chimpanzees. Anesthesia was not used. The experiments were inconclusive because the chimpanzee's skulls provided a heavier armor than human skulls. Still, the idea was to study the lapse into unconsciousness; drugs would have interfered with that process.

"Were Ms. Trull a health professional, she would have understood" that point, Moor-Jankowski wrote. Fouts provided a copy of the study directly to NBC.

NBC ended up like the men carrying the donkey in Aesop's fable.

The network wanted to please everyone and seemed unable to please anyone. The program was changed, rewritten to include more information on the positive aspects of medical research and a careful ethical debate. There were two scientists in the show—a woman friendly to the animals, a man dedicated to the goals of his profession. Animal activists thought it had become wishy-washy and primate researchers thought it was still too warm and friendly toward chimpanzees. Both sides, though, counted it as a win for researchers. It emphasized that they were learning to use their own power.

Researchers did even better, that same year, with dogs. They had been shocked by the 1991 entry on dogs that appeared in the *Encyclopedia Britannica*. It was written by Michael Fox, not the actor but a well-known advocate for animal rights. Fox, not surprisingly, did not give biomedical research an endorsement. In one part of the essay, he questioned outright the need for research "which often entails much suffering, has been questioned for its scientific validity and medical relevance to human health problems." Fox's examples of research included forcing beagles to inhale tobacco smoke and the use of dogs in military radiation studies. He also gave a plug to our society's "increasing respect for fellow creatures and growing recognition of animal rights."

Like NBC during the *Quantum Leap* fray, the publisher of *Encyclopedia Britannica* was inundated with outraged letters from biological scientists. Primate researchers waded right in. Peter Gerone wrote: "I am shocked that *Encyclopedia Britannica*, with its long history of providing unbiased educational material, would fall into the trap of becoming an instrument for dispensing false and misleading material." Researchers also announced that they would urge school libraries not to purchase the books unless future editions were free of bias.

The 1993 edition of the *Encyclopedia Britannica's* dog essay has no mention of animal rights, no mention of military experiments or smoke-breathing dogs. The president of the American Society for Pharmacology and Experimental Therapeutics wrote a congratulatory letter to *Britannica's* editor, thanking him for the changes. He also suggested another article, further explaining the important use of dogs in medical research. The society offered its services in providing possible authors for the entry, someone other than Fox.

Fox is a veterinarian, a philosopher, and to researchers a turncoat. He once supported animal research and now deplores it. In the mid-1980s, he'd actually written a book defending animal research. The book, though, was a turning point. He came under attack for his "might-makes-right" attitude. The comments were fierce enough that he began to wonder if he had overglorified the importance of humans. Slowly, he concluded that he had, that he had put himself in the position of sneering at other living beings. He now argues forcefully for animal rights, serving as director of the Center for Respect of Life and the Environment. As Fox has learned, a scientist who becomes an animal advocate enters a curious no-man's-land. He is not "one of us," to the research community, not a real scientist.

When collectors of information, historians, and librarians such as Lederer and Jacobsen are restricted, then the pressure not to stray from the pack is only that much higher for a researcher. Any professional scientist who makes the choice to work for animal welfare risks setting his career in reverse. People such as Fox or Roger Fouts know that well.

There are scientists who simply step out of the research community to work for animal advocacy. The most prominent example is Neal Barnard, who runs the Physicians Committee for Responsible Medicine. Barnard is a psychiatrist; his Washington, D.C.-based organization claims to represent about 2,500 health professionals. The group offers technical help to groups like PETA.

Many academic researchers bitterly resent scientists who speak up for animal activists. They say that some professional critics use their degrees to make their attacks on research sound respectable. Instead, scientists complain, they undermine legitimate research. They point out that many such critics no longer work in laboratories or conduct experiments. They don't do research; how can they claim to understand it? "Where do they dig these guys up from?" asks Seymour Levine at Stanford. "Not one of them has written a scientific paper. Not one of them is listed in Who's Who for their work. I'm in everybody's Who's Who. I mean, where do they find these people?"

Another question might be, why does anyone risk career suicide by even raising questions about this issue?

Stephen Kaufman sometimes wonders that himself. An Ohio physician, he is co-chairman of the Medical Research Modernization Committee, a tightly knit group of about 1,200 health professionals, co-chaired by Virginia psychologist Murry Cohen. The committee takes the position that animals are a poor research model for humans.

Kaufman lives in Cleveland, practicing ophthalmology at a local health maintenance organization. He received his M.D. from Case

Western University School of Medicine; he worked as a medical resident at New York University. He was always a science junkie, enjoyed chemistry class, actually liked math. He never doubted that he would end up in science. But, by the time he was in high school, he'd begun to be uneasy about animal research.

He didn't like the way animals were treated in the laboratory. He can provide many examples. Among the most vivid he recalls was while working at a Cleveland medical lab one summer during high school: "There was a beagle that a tech was having difficulty getting an IV line into. He kept trying and trying until the animal, obviously in pain, started shaking. It became even harder to get the line in and the tech was obviously getting frustrated and he hit the dog, hard, across the head. It struck me that the dog was being punished for being a bad research subject, something he hadn't wanted to be in the first place." Clearly, too, the animal was being punished for something else not his fault, the incompetence of the person doing the tests.

By the time he was a resident, Kaufman was also unimpressed by animal models for human disease. It's the theme of his organization. The committee wants the rest of the world to be a little more dubious, at least, about animal models. It wants people to question whether you can apply information learned in a rat to a human being. A favorite example is that of cigarette smoking. The early studies on rats were so misleading, committee documents point out, that at first researchers missed the toxic effects of the smoke entirely.

Animals simply are bad models for human health, Kaufman insists. His skepticism carries right through to primates, even chimpanzees, for all their genetic closeness to humans. Consider the AIDS virus. In humans, it kills. In chimpanzees, however close their genetics, the same virus causes a weak antibody flareup and no illness. "We are not the same," Kaufman argues, and to keep infecting and killing animals, and pretending that the match is good, amounts to both bad science and a kind of civilized cruelty.

While at NYU, Kaufman became intensely interested in such arguments. In the hours he could grab, around his ophthalmology residency, he undertook a project to analyze ten animal models, selected at random from experiments published in pathology journals. His plan was to do an objective analysis. Were the results cited by other researchers? Did clinical researchers, those treating humans, cite the animal work in their own studies?

He sought permission from NYU to do an extensive literature search, using the NYU medical library. The university turned the question over to its Animal Care and Use Committee—"which is weird, considering I wasn't planning to use a single animal," Kaufman says.

The committee turned him down flat. He ended up tacking his project together across a string of Manhattan area libraries. The analysis found that, while animal models were widely cited by other animal researchers, they appeared to be rarely as influential for those scientists working directly with humans. Further, some of the models, seemed dubious at best to Kaufman. Scientists had, for instance, been able to chemically induce colon cancer in rats. Researchers were enthusiastic about the rat model of the tumors. Yet, the great problem with human colon cancer is that it metastasizes—spreads to other parts of the body. The rat cancers simply didn't do that. They were local tumors. The comparison wasn't valid in Kaufman's opinion. He collected all of the data into an admittedly skeptical critique. Then he submitted it for publication.

The Institute for Laboratory Animal Resources, which advises the federal government on animal issues, accepted Kaufman's study for its journal, intending it as a piece for debate. The institute withdrew its acceptance though, after reviewers argued the piece was neither clear opinion nor good science. That was in 1989. The study was printed finally by Kaufman's own Medical Research Modernization Committee. Kaufman still suspects that ILAR just couldn't accept a formal criticism of animal research.

"It certainly has its frustrations," Kaufman says, a little wearily. "There's a defensiveness out there that almost borders on paranoia. Most people are smart enough not to buck the system in the first place. Because you know, if you raise these issues, you're going to get a lot of people upset. It's not a smart thing politically. I knew, by the time I got to medical school, that I wouldn't be able to go into research, that I couldn't maintain this position and have a research career."

He puts his love of research into writing carefully thought-out monographs, painstakingly criticizing animal models. Sometimes it satisfies; sometimes "I think I'm just writing academic monographs that no one reads. And it doesn't make me rich. But I think the issues are important. There are important public health issues; there are legitimate concerns about injustice toward animals, particularly when so much research can be seen as unfounded. And the other thing is that I keep my hand in out of respect for others in this fight. There are a lot of people struggling in this fight. They're good people and compassionate people. They may not be always right and neither may I, but they live by their convictions. I want to lend a hand."

Never mind, for a minute, the fights over commercial television and encyclopedias. Kaufman would just like to get members of his own profession to give a fair hearing to some of the issues involved. In 1991, he wrote to every medical school in the country. He offered to help organize a forum for physicians to discuss animal research concerns with students. Not one school even wanted to try. The dean of the University of Nevada medical school wrote: "We believe the continued use of animals in medical research and education is essential. Therefore, it is not a debatable issue."

Not debatable, not discussable, difficult even to make visible. One well-known researcher who works for animal protection is Dr. Nedim Buyukmihci (called Dr. Ned by one and all) at the University of California, Davis. In 1991, he submitted an article critical of animal research to the *American Journal of Ophthalmology*. It was rejected with this comment: "It is not entirely clear that an article taking issue with animal studies would be favorably considered by editorial referees."

In a way, it's the same dilemma faced by Larry Jacobsen. Does a journal, dedicated to a science that relies on animal research, give open space to arguments that would dismantle it? Or the decision that confronts the American Psychological Association, a group apparently unwilling to hear or publish voices of dissent from among its members. A small subset of that profession, including Roger Fouts, have stubbornly refused to be silent. In 1981, they formed Psychologists for the Ethical Treatment of Animals. The psychological society has refused to print ads from the group, including one, signed by 220 psychologists, expressing concerns about animal research. The society has also been repeatedly refused the opportunity to set up a booth in the exhibition area of the APA's conventions.

If you really want to get into hardball, though, consider the granddaddy of all these skirmishes. It took place in New York and involved Shirley McGreal, Jan Moor-Jankowski, and a libel lawsuit that spun on for more than seven years. The lawsuit was brought by an Austrian pharmaceutical firm, Immuno AG, over a (yes, another) letter from McGreal. The letter was published in a scientific journal edited by Moor-Jankowski, *The Journal of Medical Primatol-*

ogy. In the letter, McGreal criticized an Immuno plan to use African chimpanzees in a hepatitis testing program, expressing fears for the future of the animals, which would be trapped out of the wild. Immuno responded by denying all criticisms of its program, assuring officials that it would care for the chimps' future, and by filing libel lawsuits against McGreal and everyone connected with publishing the letter.

Moor-Jankowski, shocked and angry, fought the lawsuit until he prevailed. His argument, from beginning, was that all sides of an issue deserve a hearing. By the end of the litigation, though, he had come to suspect almost his whole profession of being opposed to honest debate. In all, the effect of the lawsuit has been so bitter that, although Moor-Jankowski won in the courts, you can argue that there were only losers in the fight.

To understand why this became so tough and troubling a fight, it's not enough to sift legal documents. You have to go back to the beginning, to the forests of central New York State, where Jan Moor-Jankowski helped build the Laboratory for Experimental Medicine and Surgery in Primates.

The laboratory is about an hour northwest of Manhattan in the carefully landscaped community of Sterling Forest. Research institutions cluster here—built by Union Carbide, International Paper, IBM. Just down the road from Moor-Jankowski's outpost is the laboratory of Ron Wood, also an NYU researcher, who has studied the chemistry of drug addiction in rhesus macaques, sealing the animals into small chambers and forcing them to inhale a cocaine mist. Wood has been a hot target of animal activists and Moor-Jankowski sometimes worried that they would somehow confuse his laboratory with Wood's. Or that the bad feelings would spill over to LEMSIP, which houses some 300 chimpanzees, 100 macaques, 165 marmosets, 15 baboons, and 8 squirrel monkeys.

In fact, animal advocates tend to treat LEMSIP gently. That's largely due to Moor-Jankowski. The New York primate center is not the most spacious in the country. It doesn't have the most elaborate housing. The animals are not allowed outdoors, as they are in many other facilities. Activists really like the director, though. He listens to them and he has never dismissed them as crazies. Moor-Jankowski once accepted a gift of 200 coconuts for his chimpanzees from PETA; Alex Pacheco actually appeared on NBC once, smiling and saying nice things about LEMSIP. (And Moor-Jankowski has returned the favor by writing statements in favor of PETA's attempts to reclaim the Silver Spring monkeys, arguing that the endless bad publicity does only harm to the research position.)

There is nothing fortress-like about LEMSIP—no barbed wire, no alarms, just a rambling complex of wood and concrete-block buildings, scattered through wood and meadow. Moor-Jankowski decided early on for a different kind of defense; he would be open about what he did. His laboratory would keep no secrets. He decided to make everything public, including his necropsy reports. Then, he reasoned, he would have nothing to fear from animal activists. It hasn't always worked. He once informed Shirley McGreal in advance that he wanted to bring some chimpanzees over from Holland. She contacted animal groups in Europe who effectively blocked the transfer. He still grumbles about it. But he didn't change his mind. There have been no protests at LEMSIP, no threats and attacks. When his laboratory stopped doing research with gibbons, in the 1980s, Moor-Jankowski sent seven lab animals down to McGreal's colony in South Carolina.

He has a forceful ally in his chief veterinarian and assistant director, Jim Mahoney. Not that they entirely agree. Moor-Jankowski believes in humane treatment of animals; Mahoney believes, in his heart, that animals shouldn't be in laboratories at all. Irish, thin, intense, vividly blue-eyed, and given to self-doubt, he says "You have to draw a line, and I go ahead with what I do because of what I think is a greater good, that we cannot study disease without animal models. And, yet, I don't feel so right about what I do that I could turn to an animal rights advocate, and say, you are wrong. In the end, I think *I'm* probably wrong."

Mahoney has gathered a group of animal lovers together to serve the primates there. They've been known to come in on weekends to stitch together snakes hit by cars on the New York highways. The attitude is evident on the quickest of rambles through the white-walled wings that house the animals; technicians playing songs to rhesus macaques on a xylophone, young chimpanzees watching *National Geographic* specials on a monitor, holiday decorations strung up, pink hearts dangling over the baboon cages to celebrate Valentine's Day. The chimpanzee nursery at LEMSIP is crammed with toys, play structures, and little chimps wearing diapers and Oshkosh overalls to keep the diapers in place. The humans touch and cuddle the small animals. The chimps cling and reach out and touch back, even grabbing for the fingers of passing strangers, their own leathery palms warm and quick.

It's personal for the animal caretakers at LEMSIP. Moor-Jankowski himself gives an example, told wryly. He was unhappy because wild birds had taken to nesting in the indoor eaves of some of the chimpanzee facilities; he was afraid they would spread disease. So, he decided to get rid of the birds, kill them if necessary. One of his staffers reported him to state wildlife officials.

Moor-Jankowski is not a man who laughs at himself often. But this encounter genuinely amused him. He grins when he tells it: "My own employee turned me in." He likes spirit, strong opinions, and he has an unwavering belief in the right of each person to speak his or her mind. For himself, it's simply people first. If the chimpanzee nursery bears Mahoney's stamp, then the quarters for the adult chimps are more clearly Moor-Jankowski's plans. They are built to keep the animals healthy; cages hung from the ceiling so that feces can drip through mesh bottoms onto plastic liners, removed every day. The chimps do not go outside where wild birds and animals track viruses around. By Moor-Jankowski's standards, his first duty to his animals is to keep them healthy. Not to cuddle up with them.

"I really love people," Moor-Jankowski says. "I like to be with them. I cannot sublimate with animals. I can't play with a dog. It's not a companion. That's why I went into primate research in the beginning. I wanted to study man—but not experiment on people. Things were different then, it was very utilitarian with animals."

He looks the effortless aristocrat; tall, slim, high cheekbones, brown eyes behind gold-rimmed glasses, a charcoal gray suit with matching vest, blue shirt with fine white stripes, a tie patterned red over blue. There's a story floating around the primate community that he once got into an argument with a veterinarian, who finally summed up their differences by saying "Well, MJ, I guess I'm a blue jeans kind of guy and you're a Brooks Brothers kind of guy." Moor-Jankowski, the story goes, replied, with some irritation, that he did not buy suits off the rack.

Among the notices and documents stacked over his desk at New York University is a reminder from a Hong Kong tailor. Moor-Jankowski lives in a Manhattan apartment overlooking the silvery swath of the Hudson River. He drives a Mercedes sedan. But it is all polish, in a sense, over a seemingly bottomless sense of honor, of what's right, and an unwavering determination to hold to that. He won't tell the stories, but he was Polish born, fought in the underground resistance during World War II, fiercely refusing to give in to Nazi dominance.

During the Immuno lawsuit trial, he engaged in an angry exchange with the pharmaceutical company's lawyers, accusing them of badgering him, of trying to trick him with repeated variations on the same question." I've had enough of this Gestapo type of question," he snapped. "I was exposed to it once in my life." He then stomped out of the room, leaving the Immuno lawyer to complain that "Dr. Moor-Jankowski seems to be having a problem."

Moor-Jankowski came to the United States in 1963 from Great Britain, where he had built a reputation studying blood types in baboons. He worked at NIH, but was not happy. "I hated it," he recalls. "They all lock their doors there so no one can steal their ideas." He moved to the Yerkes center, and then was recruited to move to New York and help establish a primate center. The New York research community had begun to feel excluded. It was a point of pride. Why should there be federal primate centers in Wisconsin and Oregon and none in New York? When LEMSIP was originally founded—Moor-Jankowski became director in 1965—the idea was to take it to the federal government as an already organized facility, and become the eighth NIH center. Other primate centers opposed it forcefully; Moor-Jankowski believes they just didn't want to share the money. At one point, New York legislators actually got a bill through Congress, requiring that LEMSIP be made a federal center. President Reagan let the bill lapse. With that, New York gave up the fight.

The politics did not improve relations between Moor-Jankowski and the National Institutes of Health. They were already strained. In 1974, NIH had given the New York center a contract for a breeding colony of chimpanzees. Four years later, though, it switched the contract, shipping all 71 animals to the Southwest Foundation for Biomedical Research in Texas. Moor-Jankowski felt betrayed, especially since the colony had been very successful. He persuaded NYU to sue over the switch. The lawsuit failed and also angered federal authorities. The disappearance of that grant, almost $700,000 a year out of a $1.7 million budget, almost bankrupt the lab. The university considered closing it down.

Moor-Jankowski, though, had international connections. He was able to begin contract work for private companies, first in Japan, then in Europe. Today LEMSIP remains an independent facility, run on about $4.5 million a year. It works closely with universities and industries around the world. Yet Moor-Jankowski's relationship with NIH has remained unfriendly. Moor-Jankowski rarely gives up

on a fight or a perceived enemy. Since NIH took away his chimpanzee breeding program, he has not tried to improve relations. He has
become instead publicly critical of the government's primate work.
The letters in support of PETA in the Silver Spring monkey fight
are an example of that. In response, he says, NIH has done its best
to act as if he doesn't exist.

He points out that LEMSIP has one of the best primate blood-
typing operations in the world, that it is in fact designated by the
World Health Organization as a center for that purpose. Yet in
1984, in fact California surgeon Leonard Bailey was planning to
transplant a baboon heart into a baby and raising questions about
primate blood types, federal health officials did not refer him to
LEMSIP. The operation failed, in part because the baby and the baboon had different blood types; the proteins in the animal's heart
caused the baby's blood system to rebel against it. Bailey was put in
touch with LEMSIP only after the failure. Moor-Jankowski says his
lab had already documented such problems in mixing blood types
during transplants. He believes that if Bailey had known about
LEMSIP's work, he might have been able to save the baby.

"NIH has ostracized our results, put us under quarantine," Moor-
Jankowski complains. No one, of course, could make LEMSIP an
invisible laboratory, especially with such a visible director. It has
attracted some of the top primate virologists in the country, among
them Elizabeth Muchmore and Preston Marx. The research team
there helped pioneer the first hepatitis B vaccine through work with
chimpanzees; helped develop techniques for freezing blood for storage; has done some remarkable work on the chemistry of pregnancy
that has attracted the attention of drug companies worldwide.

Moor-Jankowski continues to edit the respectable *Journal of Medical Primatology*. And for all his unhappiness with NIH, Moor-
Jankowski has remained a fully open-minded editor. Many times,
the journal's lead author will be from NIH, writing on some topic
purely of interest to primatologists, such as "Diseases of the Callitrichidae [marmosets]: A Review." Letters to the Editor tend to be
equally technical in nature, bearing such titles as "Assessment of
Semen Quality in a Baboon Breeding Colony."

That may explain, in part, why Shirley McGreal's letter, in December 1983, seemed to attract so much unfriendly attention, to
stand out like a chain-smoker at a convention of the American
Lung Association.

Of course, the headline in a journal with a tendency to use words like Callitrichidae was something of a screamer: "A Project with Potential to spread Non-A/Non-B Hepatitis in West Africa." Non-A/Non-B hepatitis is also known as hepatitis C. The five hepatitis viruses, which attack the liver, go by the no-nonsense names of A through E. The first three—A, B, and C (or Non-A/Non-B)—have proved the most troubling. But there are measures to deal with A and B, such as sanitation, drugs, and vaccines. The others have proved tougher. Immuno wanted to study Non-A/Non-B in chimpanzees, working toward a preventive vaccine.

The company's tentative idea was to set up a research station in Sierra Leone, with a natural population of the animals. McGreal's letter made it clear that she didn't like the idea. She didn't argue the merits of wiping out hepatitis or even using chimpanzees. Her concern was that, by setting up in Sierra Leone, the company would use wild chimpanzees, an endangered species. Theoretically, the animals could be studied and then re-released, since hepatitis is not a troubling illness in the apes. Immuno had indicated that it did not want to keep the chimpanzees permanently captive. McGreal listed four concerns: (1) that the track record for re-release in general had been dismal; (2) that the chimps might infect other animals; (3) that they were desperately endangered in the wild, and (4) that there was already a supply of captive chimpanzees in the United States and other countries. Why not use those?

Her opinion was carefully worded to the point of stuffiness. She closed the letter like this: "The International Primate Protection League shares the scientific community's concern over hepatitis. However, we feel that a way can and must be found to resolve this problem without recourse to the dwindling populations of wild chimpanzees. Therefore, we appreciate the opportunity to draw this situation to the attention of interested parties."

Before publication, Moor-Jankowski sent a copy of the letter to Immuno, offering the company a chance to answer her concerns. He knew Immuno. His laboratory had done contract work for the company before. Immuno, though, would not respond to the charges; the company's lawyers made it clear that the letter was not welcome. Moor-Jankowski waited almost a year, then published McGreal's letter in December 1983. By that time, he himself had become somewhat critical of the plan; he was quoted in the British journal, *New Scientist*, as calling the move into Africa "scientific imperialism."

To say that Immuno didn't appreciate being made a public target is an understatement. The company filed defamation lawsuits against McGreal, Moor-Jankowski, the *New Scientist*, the *New Scientist* writer, the *Journal of Medical Primatology*, New York University, and the distributors of both publications. It asked $4 million in defamation damages from each, insisting that the criticisms were unfair, untrue, and overstated both Immuno's plans and the problems attached to them, and understated the company's commitment to protecting wildlife. As one of the firm's attorneys explained carefully to *The New York Press:* "We weren't looking to shut anybody down or make anybody knuckle under to pressure. We were just trying to clear our name."

It was clear, almost from the beginning, that Immuno's lawyers were prepared to play tough. In one notable deposition session with McGreal, Immuno lawyer Raymond Fersko began softly, asking her if she worked for conservation when she attended international meetings on endangered species. McGreal acknowledged it. The deposition then rolled on like this:

FERSKO, TO MCGREAL: Did you ever perform any sexual acts at any of these conventions to try to persuade delegates to vote a certain way?"

HENRY KAUFMAN *(McGreal's lawyer):*   You've got to be kidding. I am going to move to strike that question.

MCGREAL: Where did you get that from?

FERSKO, TO KAUFMAN: Do you direct her not to answer the question?

KAUFMAN *(after the attorneys agree to keep the question on the record):*   They must really be paying you a lot for this one.

Eventually, Immuno decided not to pursue the chimpanzee experiments. But it was unforgiving on the issue of defamation; the company stuck the lawsuit out for some seven years. In the face of such determination, everyone settled except Moor-Jankowski, McGreal's insurance company bailing out over her angry protests. She refused to sign the settlement in protest; her insurance company dropped her as a client.

Moor-Jankowski, who had fought with the Polish resistance, couldn't make himself quit. His own insurance company begged him to settle; it threatened to cut off payments if he kept on. He sued the insurance company and won. He was tired and worried and in debt; the work at LEMSIP was suffering because his concentration was all in the courtroom. There was still that ironclad sense of right and wrong, and this, to him, was wrong.

"Jan was extraordinary in that he fought it out," says his attorney Philip Byler, with the Manhattan firm of Stults Balber Horton & Slotnick. "There's an eight-volume legal record that cost $16,000 just to print. There were $2 million in legal costs. (Moor-Jankowski paid $70,000 out of his own pocket because depositions were required in Europe and his insurer didn't cover the fees of foreign lawyers.) How many times does that result from a letter to the editor?"

The first ruling, from the New York County Supreme Court, went against Moor-Jankowski. But the New York Court of Appeals overturned that verdict in a blistering and unanimous reversal, calling the case "an ill-focused libel suit" without one actionable statement. The court reviewed 4,000 pages of testimony and documents before concluding that McGreal's letter was a combination of solid fact and simple opinion, both protected under the First Amendment. The case rocked around for some time yet. The highest state appeals court upheld; Immuno appealed to the U.S. Supreme Court; the U.S. court sent it back; the New York court upheld again; and the U.S. Supreme Court refused to rehear the case. That ended the litigation.

Philip Byler is a passionate defender of free speech, a savvy First Amendment lawyer. From the beginning, Moor-Jankowski's case has been straightforward to him—simply concerning the right to express an opinion. McGreal's letter was factual, he says. That's not required for a letter to the editor. The letter is about your opinion. You have a right, Byler thinks, to be wrong, to be weird, to make a fool of yourself. Not every letter writer should be held to the standard, even, that McGreal held herself, he argues.

"It was two pages, based on factual statements, all of which are correct," he says, flatly. Byler is fair-haired and blue-eyed. His voice is quick and sure. "It was a very careful letter; she even took it down to scientists at the University of South Carolina to check it out. And is that the standard we should hold for people who write letters to the editor? We're talking about the flow of public opinion. Should people be denied the chance to express their opinion because a very wealthy corporation doesn't like the debate?"

Moor-Jankowski had a lot of supporters by the time the lawsuit ran itself out. There were universities: Barnard College, Syracuse

University, and Fordham University. There were environmentalists: the Sierra Club, the National Audubon Society, the World Wildlife Fund. There was the media out in force: CBS, ABC, and NBC; *Newsday*, *The New York Times*, the *New York Daily News*, Time Inc., the Association of American Publishers, the Magazine Publishers of America, Inc. Not surprisingly, the animal protection movement came in: the Humane Society of the United States, the Animal Welfare Institute, Fund for Animals, Friends of Animals, the American Society for the Prevention of Cruelty to Animals. Thirty-four groups backed him up in court.

The list didn't include one of the big science or medical societies. Not the American Medical Association, the Society for Neuroscience, the American Association for the Advancement of Science. They were invisible—or they were against him. Perhaps if he had a smoother relationship with his colleagues, if he had not wrestled so visibly with NIH, if he had not been friendly with PETA, the science community would have stood behind him. And perhaps—considering the bitterness over animal activists, the hostility toward Shirley McGreal—they wouldn't have helped him anyway.

In fact, if you consider the National Association for Biomedical Research to be representative, then you can only conclude that researchers did not support unfettered freedom of expression, at least not for everyone. NABR also filed supporting briefs in the lawsuit, both times that the New York Court of Appeals considered the case. It did not take the side of Moor-Jankowski or New York University, which happens to be a member of the association. It swung with Immuno, over the protests of NYU and the disbelieving anger of Moor-Jankowski. Once was upsetting. The second time, in Moor-Jankowski's eyes, was vengeful; he believed that the biomedical community somehow wanted him to pay for not behaving like an insider.

The associate dean of the NYU medical center engaged in a dismayed correspondence with Frankie Trull at NABR. Dr. David Scotch insisted that NABR's position was "ill advised and contrary to the best interests of at least the academic members of the Association." Trull wrote back that NABR was not trying to support either party, merely providing pertinent information. Scotch argued again: "Contrary to your statement that the Association does not intend to support either party in the litigation, the attorneys representing the Association have explicitly stated their desire to file a brief in support of the plaintiff" (Immuno). He urged that the board be made aware of NYU's position and reconsider "the wisdom" of essentially slapping down one its own members.

Trull's written reply pointed out that members, such as NYU, elected the board to make decisions for them. She reassured Scotch that the association was all for freedom of speech. But, "As you know, animal research is under constant attack. . . . While we are not interested in the outcome of this case as it affects the plaintiff, we are most interested in preventing damage to research while being consistent with First Amendment principles." In frustration, NYU filed another brief of its own with the court, publicly announcing its lack of agreement with the stand taken by its own association.

To understand why there was such a strong reaction to the NABR brief, you have to appreciate that it took a very tough approach. The brief stated, in part: "The antivivisectionists . . . have waged a vicious guerrilla campaign that consistently includes resorting to deliberate dissemination of whole lies, half-truths and innuedo. Moreover, the antivivisectionists have not stopped at mere verbal violence. They have engaged in arson, death threats, bomb threats, vandalism, theft, break-ins, obscene phone calls, intrusive picketing and abusive harassment of individual scientists. . . . These scurrilous tactics of disinformation, malicious mayhem and brute intimidation have unlawfully undermined rather than enriched or enlivened the marketplace of ideas."

Further, the brief states, laboratory directors, such as Moor-Jankowski, are fully informed of such tactics. "Defendent-respondent Moor-Jankowski, as the record indicates, was very aware of those antivivisectionist activities and the grave threat those activities pose to research." Because of that, the NABR attorneys argued that Moor-Jankowski should have known better than to allow McGreal to publish a letter in his journal. He should have recognized, the brief says, that the letter would be loaded with inflammatory falsehoods. He had a responsibility to verify all her facts. A medical journal was no place for an animal advocate to advance her cause. Editors of such publications, NABR argued, should be held to a higher standard than, for instance, the editors of newspapers and magazines. Newspaper editors might allow a factually sloppy letter to the editor slide through; scientists could not. The NABR brief emphasized that the organization made the argument because of its commitment to "truthfulness to the presentation of and reporting upon the research activities of its membership."

The brief concludes: "It's important to keep in mind that, by ruling in Immuno's favor, this Court will not in any way diminish New York's preeminent status as a protector of First Amendment freedoms and as a bastion of scientific ferment." A pro-Immuno ruling would, instead, send a signal to the editors of scientific journals that they would have to be meticulous in the letters they allowed to be printed, that theirs was no space to "further the cause of a particular pressure group."

By this time, Byler was almost as infuriated as Moor-Jankowski. He filed a heated affidavit in response to the NABR brief. First, on the principle of free speech, they would not oppose the association entering its opinion. Nevertheless, he wrote, the opinion was irresponsible, less than candid, and "an attempt to justify the misuse of libel laws as a tool of oppression against a political view not to the liking of the National Association for Biomedical Research—wildlife conservation." Also, Byler complained, the association was using the lawsuit to simply attack animal protection groups, which had nothing to do with the legal matter at hand.

"Indeed, what is so ironic about NABR's brief is that it purports to profess the virtues of 'civility,' 'morality,' and 'truthfulness' and yet it engages in a broadside political smear against animal conservationists in the context of this case, where the Letter to the Editor was so clearly civil, moral, and truthful in expressing conservationist 'concerns.' "

Further, Byler continued, there was nothing wrong in a medical journal airing a variety of opinions about research. He quoted from one of Moor-Jankowski's answers during deposition time, when Immuno lawyers asked the lab director if he considered that McGreal acted in a "conventional" manner. Moor-Jankowski had replied: "The last time I experienced people trying to judge somebody, wanting everybody to be conventional, was in Nazi Germany. This is a mark of a totalitarian system, to want everybody to be the same and conventional."

The New York justices agreed. And from that point on, legally, everything went Moor-Jankowski's way. Yet, the really triumphant wins are the clean ones, the ones where the decision comes and you celebrate and you go on. You can leave the hurt and the anger behind. This hasn't been like that. You can sit with Moor-Jankowski now, in his gold-tone Mercedes, with the lights of Manhattan gleaming through the windshield; in his paper-filled office at NYU. The anger still seeps through. The hurt is still there and he still feels betrayed by his own. Where was the support from the research community; why weren't they there? He can't change that, but he can't let it go either; letting it go would be like letting down his defense and, he says, "I am not easy prey."

His bitterness against NABR remains unabated. He challenges Frankie Trull now, where he can. He believes that her organization has worked to make the animal research issue worse, that it has helped drive the polarization over animal rights. She is "pernicious and nefarious; she is divisive." He can drive Trull into an unusual defensiveness and he does.

"Because I'm president of this association, I get the rap for everything, including that lawsuit," she says. "I didn't make that decision. It was the board's decision. If it was entirely up to me—who wants to get involved in a lawsuit? But the board felt there was so much misinformation being circulated, that there needed to be a burden of proof on the publisher, that whoever publishes needs to be accountable. It wasn't an easy decision. But our lawyers made the presentation, with all the pros and cons. And our lawyers still think it was the right decision. I know Dr. Moor-Jankowski in some ways feels badly, that this was a mission against him. He just won't let it go. He takes shots at me whenever he can. And I don't think it's healthy. He should declare victory and go home."

Moor-Jankowski had friends who told him that NIH had provided information to Immuno. NIH officials flatly deny it. Moor-Jankowski doesn't believe them. He's mistrustful enough that when federal health officials were collecting data for their chimpanzee conservation program and wanted to know how many chimps live at LEM-SIP, the laboratory refused to provide the information. "He calls us uncooperative," complains one NIH official. "He wouldn't even give us a simple number." What would they do with it, Moor-Jankowski asks, what enemy of his would they leak it to? He's still paying off his $70,000 debt, refocusing on his laboratory, and he's tired. There are days when he thinks he has paid too high a price in defending his laboratory. "When I came here, you know, I was Mr. Primate. But if I was starting my career now, I wouldn't work with animals. I've thought about some of the things scientists do to animals. And I think they are wrong."

Byler is not surprised that Moor-Jankowski is suffering from burnout. It was a long siege and, he thinks, a frightening one: "What happened, as a result of the Immuno case, was not as bad as it could have been, to an extent, because of whom Jan and Shirley are. They weathered it. But the average person, if he had an idea that a letter to editor would lead to a very expensive lawsuit, might not have stood it out. Remember the old Jim Crow juries? It was very difficult to survive in that atmosphere of suppression. And this is about the same kinds of repressive attitudes." And on those grounds, he believes, Moor-Jankowski would do well to remain in a defensive posture.

Byler's impressions are not so different from Susan Lederer's. Both acknowledge that the research community, if it chooses, can retreat into secrecy, shut down debate. Both believe that such a course will do harm. Retreating may provide shelter. But in the long run, Phil Byler will argue, secrecy makes for a damned leaky roof: "Ultimately, it's dangerous for scientists not to debate these issues and it is self-defeating. Scientists need to be open and sensitive to issues like conservation. And if they refuse to engage in the debate, they are kidding themselves if they think that some people aren't going to perceive that they are hiding because their research is frivolous. They can do more harm to themselves that way than anyone else."

EIGHT

The Salt in the Soup

W<small>HEN JEFF ROBERTS</small> was 24, his boss locked him in a gorilla cage and walked away.

Roberts is now in his forties, assistant director of one of the largest primate research centers in the country, at the University of California, Davis. He works surrounded by monkeys in cages. In the outdoor corrals and pens, in field cages spanning half an acre each, in the neat racks of indoor cages, there are 2,703 rhesus macaques, 492 crab-eating macaques, 160 squirrel monkeys, 60 bonnet macaques, and 42 titi monkeys.

On a summer day, in the dry heat and rough greenery of California's Central Valley, standing on the pebbly ground of the primate center, you can shut your eyes and imagine that you are in a jungle, buffeted by the rush of wind and the chatter of monkeys. Rhesus macaques are talkers, cage-rattlers, bar-bangers, vocal about their fears, likes, dislikes. They are so noisy that, when housed indoors, labs will often try to buffer other, less exuberantly talky species from their chatter.

Roberts doesn't let himself slip into fancy. Over the rowdy squabbling of a troop of macaques, he still hears the nearby hum of Interstate-80 as it stretches southwest across the valley toward San Francisco. The greenery around him is not wild but landscaped, the fig trees neatly arranged in clusters near the big cages. The California Regional Primate Research Center boasts 17 half-acre field cages, bustling with macaque society; young females baby-sitting their siblings, sulky groups of adolescents, females grooming the big, tough males in a gently seductive manner. But Roberts knows that at its edges, a cage is still a trap.

It's been almost 20 years since he stood in the cage at Chicago's

Brookfield Zoo. With a new degree in biology, he was the youngest staffer in the primate house at the old urban zoo. He retains the restless energy he had then, the very clear silver-blue eyes. He's stockier now, maybe, and his dark hair is shading to gray, but he still has a grin full of mischief and he still likes monkeys a lot. When he started at the zoo, he'd been given a choice between small animals and primates. He picked primates because he was curious about them and because he thought the caretakers were an interesting group, fiercely loyal to the animals they worked with.

"They were the armpit of the zoo," he recalls, that mischievous smile lighting his face. "When the director hired me, he said it was like pitting 'good' against 'evil.' They were considered the bad neighborhood in the zoo, people with an attitude. A couple years later, he told me that 'evil' had won out." The director believed that, like his coworkers, Roberts had been seduced by the monkeys and apes. There were round-eyed lemurs with their coiling tails; long-nosed baboons; quick-thinking African guenons; spider monkeys, as fast and fragile as dandelion fluff; long-armed orangutans; gorillas, black and broad. He could sit in his spare moments and play footsie with a gorilla, marvel at the gentleness of the big ape's touch. "It was the best job I ever had," he says. "Here, you can't really capture that experience of going one on one with the animals. But back then, I'd go on my day off just to play with them. They were that incredible."

He'd just started at Brookfield when the head keeper locked him up for the afternoon. The zoo had a series of interlinking holding cages for the big orangutans and gorillas, where they could keep the animals while the main cages were cleaned. The holding pens were boxes, iron-barred, concrete-floored. "He asked me to step inside one of the cages for a minute. And then he locked it. He told me that he'd be back in a few hours, that he wanted me to think about what it was like to be inside. After you sit there about an hour, you're peeking around the corner, wondering if a keeper is coming, wondering it he'll bring you water. It's a helpless feeling. You are totally dependent on someone else."

More than that, he remembers the boredom—minutes dragging along while he waited to be let out, back among people, back among friends. It's not like that for caged monkeys at a research laboratory. Life in a box may be boring, the cage may be the trap, but it is also safety, a sanctuary. Being taken out means an experiment or some medical treatment. It means being handled by humans. If the monkey is housed with companions, leaving the cage means being separated from its own and pulled into a threatening outside world.

Monkeys, even the most experienced research animals, go fearfully into the human world. At Roberts' center, scientists showed how nervous the animals were. They took six female rhesus macaques and implanted radio transmitters beneath their skin, near the center of their chests. The transmitters broadcast the pace of the monkey's hearts: when the beats skittered nervously, and when they settled into a steady thud of calm. To the surprise of the researchers, even well-known routine practices—cleaning cages, testing for tuberculosis—pushed the macaque's heart rates into a skidding pace. Even several hours after the tests, the monkeys' heart rates were still hurrying.

The listening technique is called telemetry. Without it, you wouldn't know about those racing hearts. More important than the technology, though, you wouldn't have that answer if you didn't care to ask the question. You have to care about the heart rate of macaques. You have to wonder what it's like to be a monkey in a cage. You have to be willing, even for a minute, to put yourself in the animal's place, to be Jeff Roberts, in 1974, standing behind bars in the Brookfield Zoo.

"I've never tried that on anyone out here," he says. "The single cages are too small, you couldn't get a person in. And the field cages—half an acre, it's not the same. But I use it as an example. I use it as an example of the dependency that animals have on caretakers. Say you're trying to train an animal. And it's not going the way you want and your temper starts rising. I say, go outside and think about what that animal's situation is and what your situation is. You're the one who can go outside. So make it as easy a process as you can for you and the animal."

Animal activists complain that scientists don't give them a fair hearing. On this ground, the reverse is true. Activists rarely give fair credit to the researcher who works quietly within the system, even though many have brought about real change in the country's laboratories. This falls again into the category of telling only part of the truth. If you spend all your time talking about what Martin Stephens, of the Humane Society of the United States, calls "the bad apples," then you leave no room for the rest of the crop—the veterinarians, the psychologists, the behaviorists, the animal technicians, the individuals within the community who believe that status quo is not good enough. The often-cited "research community" is not made up of one single-minded scientist, cloned multiple times.

At Davis, Roberts has gone head-on with scientists who oppose or disregard the firm standards he sets for taking care of monkeys. In one memorable instance, a nationally known researcher, specializing in fetal medicine, drove from Nevada to observe a surgery. The doctor was refused entrance to the operating room, on Roberts's orders, because he had failed, despite several requests, to get a tuberculosis test. Tuberculosis is a feared disease in primate research centers, capable of infecting a roomful of monkeys within days. The day of the surgery, the scientist stood kicking the furniture in a waiting room, complaining that it was easier to get in on a human surgery. There he stayed.

"It struck me, after being in that cage, that the animals needed an agent, someone to speak for them," Roberts says. "That I could tell people what it was like, how it made me feel to be locked in the cage. And that they couldn't."

These days Roberts's viewpoint has an additional power, beyond what he could achieve alone. The U.S. government requires scientists to put themselves in the monkey's place—in the cage, as it were. The rule is part of the controversial 1985 revisions to the animal welfare act, the section that says if you run a laboratory in this country today, you are legally responsible for the "psychological well-being of primates."

Those few words have tied the science community in knots since they appeared. They tied USDA up, too. The agency spent almost three years wrestling with that question. The agency held conferences, it called in experts—including the 1987 panel that so frustrated Roger Fouts. It sought advice from around the country, trying to figure out how its inspectors could assure the mental well-being of a monkey. After all, USDA inspectors do not have time to evaluate every animal in detail. Consider the agency's huge western division, responsible for 13 states. There are 18 veterinarians assigned to inspect nearly 1,000 research institutions, zoos, circuses, and performing animal acts. They carry out between 2,500 and 3,000 site visits a year. The idea that they would add to their burden the responsibility for evaluating the mental health of monkeys was almost mind-boggling.

The head of the western sector, Ron DeHaven, himself a veterinarian, has developed a determined cheerfulness about the law. De-Haven says the act forced the agency to change the way it did business, for the better. Previously, USDA inspectors were essentially building inspectors. They looked for peeling paint, dirty cages, proper provisions of water and food. "Now we're making the transition to being animal welfare inspectors," he says. "And I think it's the next logical step, to consider the psychological status of the animals. This is a valid issue. The days where you put a nonhuman primate in a square, stainless steel cage and leave him there for 5 to 10 years, those days are over. And they need to be."

The act, of course, is not just about monkeys. It covers vertebrate animals in captivity—primates, dogs, cats, rabbits, guinea pigs, whales. It does not include rats and mice, although animal advocacy groups have filed suit against that exclusion. There are other parts of the act that researchers and activists have fought over: a requirement for daily exercise of laboratory dogs, and what the standard cage size should be. Those are at least straightforward issues, measurable by timeclocks and tape measures. To be fair, exercising a dog may actually improve its "psychological well-being." Still, the act does not require that a scientist get nose-to-nose with a dog and find out if it's having a good time. It is only with monkeys and apes that such a demand is made.

It should come as no surprise that the USDA has yet to come up with even a formal definition of what psychological well-being means. It should also be no surprise that animal activists and researchers do not agree on what's needed. USDA—with strong encouragement from the biomedical research community—took what you might call a flexible approach. Each institution had to come up with a plan to address psychological well-being. The plan had to take into account key issues: that monkeys are social; that they are bored in an empty cage; that young and old need special care; that the government was dubious about using the animals in multiple surgeries or in experiments requiring that they be tied down or locked into a restraining chair.

USDA raised those issues in the regulations but it did not set one standard for all labs to follow in all cases. Animal activists hate that flexibility. The regulations allow each institution to draft its own plan, based on the types of monkeys it housed, the kind of experiments done. There is no absolute way to measure success or failure. Cages do not have to be larger; there's no absolute requirement for certain toys, no requirement even for group housing. The approach is called "performance standards"; it allows room for interpretation, creativity and, therefore, great variety.

Activists want "engineering standards" that mean tough, clear regulations. They want a report card; the legal battles over the Animal Welfare Act turn largely on that issue. Animal advocates want a guarantee that monkeys' lives will be improved—larger cages, better food. The USDA regulations essentially trust the institutions to do right. If you concentrate on the bad-apple researchers, there's nothing in the track record to warrant such trust. Nothing, for instance, in the Silver Spring debacle makes Alex Pacheco think researchers care about monkeys' psychological well-being. Christine Stevens, who helped lead the legal fight against USDA's "performance standards," complained that they simply "gave it all back to the researchers and said, 'Here, do what you want.' "

Yet, if USDA were to set a hard rule, what would that standard be? There are about 250 species of primates in the world, perhaps a tenth of them are used in biomedical research. They are all different. If a rat does not equal a pig does not equal a boy, then neither does a macaque equal a squirrel monkey equal a baboon.

Squirrel monkeys are a fiercely feminist society. In the wild, the females hang together—the inner circle—and the males hover at the edges, permitted in only during the mating season. Put a male squirrel monkey into a cage full of females, and you are guaranteeing him misery. "They will beat the hell out of him," says Stanford's Seymour Levine. Rhesus macaques live within a rigid and intolerant caste system that has less to do with sex than with the family one is born into. Male and female is not the issue here, it's the monkey version of feudal society, from a pampered royal family on down to a class of servile slum-dwellers. Baboons are a patriarchal society, dominated by males and fascinated by food. Hunting for the daily meal is one of their favorite occupations; they are both consistent and gracious about it. Biologists in Africa tell tales of visiting baboon troops and watching them carefully allow the humans foraging room.

To treat all monkeys as alike would be naive. Yet, barring years of study—which activists regard as a delaying tactic—how do you find out what each species needs? We are not yet at a point where a researcher can amble up to a cage and say, "So, how do you like it in there?" Even with chimpanzees, those most sophisticated apes, scientists haven't achieved that kind of communication. You might guess that a boxy little cage wasn't a first choice, but asking that question is another matter. Even when you know the animals well, you can still guess wrong.

Up in Ellensburg, Washington, on their shoestring budget, it took Roger and Deborah Fouts 13 years to raise the money for chimpanzee housing that included outdoor cages. Since Fouts moved there in 1980, the chimps had been confined on the fourth floor of the psychology building. They had lots of ramble room, mesh tunnels connecting big-barred cages. They couldn't get outside, though. At best, they had a window view of trees and sky and nothing more.

When they moved into their new building, in May 1993, the Foutses expected the chimps to hate it at first. They thought it would be too much of a culture shock. Suddenly, the chimpanzees had 7,000 square feet of living space, indoor-outdoor pens. The outdoor area soared 32 feet up; old fire hoses had been woven to make climbing structures for the chimpanzees. It was such a different world, they thought it would frighten the animals. Instead, the apes seemed thrilled. They were livelier, less aggressive with each other. "Play time has increased sevenfold," says Roger Fouts. "I've started to wonder if we're going about measuring psychological well-being wrong. We should look at how much the animals play. You play when you're happy."

Deborah Fouts recalls that when the door opened and sunlight flooded into the room, Tatu first sat staring and then began screaming and hand-signing, "Hug, hug, hug," before bolting through the door. The Foutses were determined to give the chimpanzees at least mornings outside. They even opened the door on rainy mornings. In the dripping rain, Moja sat grouchily within an outdoor structure. But Tatu climbed halfway up the cage, to a ledge, and just sat there, water running off her ears, dripping into her eyes. She signed quietly to herself, "Out, out, out."

There's not a rhesus monkey today—not a capuchin, a squirrel monkey, a baboon—that can communicate on that level. The closest anyone has come to that breakthrough with monkeys is Duane Rumbaugh, with his computer game stars, Abel and Baker. Rumbaugh has plans to teach them human language, working with the same symbols his chimps have used. If they ever did learn to tap out a simple sentence, what would they say? Would they ask for more room, as Roger Fouts might suspect?

Some scientists look at those federally mandated words—"psychological well-being of primates"—and wish for an easier goal, say, an exploration of the outer planets. No one feels those frustrations more these days than Irwin Bernstein, a Brooklyn-born, University of Georgia psychologist who just happens to head up the "Psychological Well-Being of Primates" Task Force for the National Research Council. The task force, on which Duane Rumbaugh also serves, is supposed to provide recommendations to the rest of the research community.

Bernstein turned the job down flat when first offered it. He was pressured into it, he says, and he might still have refused, except that the research council has promised not to print the names of the authors on the report. The final report will simply come under the name of a division of the research council, the Institute for Laboratory Animal Resources. He consented to take the job for another reason, too: Like Jeff Roberts, Bernstein likes monkeys.

"We'll probably be condemned by everyone, in which case we'll have done our job well," Bernstein says, ruefully, in his sunlit office on the Athens, Georgia campus, a room filled with books and photos of monkeys. He has invited many people to present their opinions to his committee, including Roger Fouts and Christine Stevens. But he has made a point of keeping activists—from either side—off the committee. "A lot of people were offended that their favorite spokesman wasn't part of our group. But their favorite spokesman had already taken a public position that he couldn't back away from. We didn't want animal rightists, and we didn't want people who would come in raving about biomedical research. Someone like Alex Pacheco, he's so visible, he's so high-profile. He does represent a lot of people, but I don't think we could have a useful dialogue."

On the question of monkey mental health, Bernstein also wrote John Melcher, the former U.S. senator who inserted the "psychological well-being" clause into the Animal Welfare Act. The letter asked Melcher what he meant by it. "He wrote us back and said [that] to him, it meant larger cages. Well, consider this—rats like little places. You put a rat into a nice, big open field and it is terrified. A big house is not necessarily more desirable than a small one."

The basic cage standard today is that a monkey, housed alone, must have room equal to at least three times its standing area. In human terms, basically a closet. Most monkeys actually get a lot more room than that; at NASA-Ames in Mountain View, California, the animals prowl around old dog runs, some twenty or more times their size; at the big primate centers, especially in the mild Southern climate, many monkeys are kept in multiacre enclosures.

Yet, the group at the big California primate center, again using telemetry, has not found that bigger cages produce any noticeable signs of relaxation, a slowed heart rate for instance.

Bernstein complains that people keep thinking of cage size the way they think of comfortable space for humans, square footage in an apartment. How much walking room do you have? Three times body size is equal to three maybe four steps in any direction. Fine for a monkey who likes to sit, even for slow walkers. But what does that do for monkeys who love to run, like crab-eating macaques, born sprinters? Or consider gibbons, those consummate aerialists of the ape family, who move by swinging hand over hand. One swing by a gibbon can take it 5 feet; a 10-foot square cage would allow it only a couple of strides. Many monkeys, too, are climbers. Floor space means nothing to them. They want to go up.

"We did one study to find out if animals are stressed in a small cage," Bernstein said. "We put them in small cages and then we got up close and stared at them. And they just did continuous back flips and bizarre running motions. In a medium-sized cage, there was less of that. And if we put them in a large cage, they just ran away. We concluded that [the behavior] was a thwarted escape attempt. They were trying to run away and they couldn't. So they'd lock into a kind of locomotor pattern. Given adequate room, they escaped. But it's not just the size of the cage. It's also the design of the cage itself."

Bernstein tells a story of animal behaviorists, working with caged tigers at a zoo. Many of the caged big cats were not taking care of their young. The researchers put a wooden partition into the cages, screening the animals from the visitors crowded around them. The female tigers immediately shifted into motherhood, apparently just needing a place of escape and privacy from staring humans. "I'm not saying we need 50-foot cages," Bernstein says. "But maybe a place to hide."

It is on this point that USDA administrators, such as DeHaven, find themselves hung. If they are required to set one standard for all, there's no doubt that they will make some monkeys miserable. There's no subtlety to a law. In the matter of hygiene, for instance, everyone agrees that a clean cage is a well-cared for cage. Yet, consider the problem of marmosets. They are vertical climbers. And, like cats, they are territory markers, carefully marking their space with urine. Keep the cages scrubbed to meet USDA sterility requirements and you can end up with a confused, neurotic marmoset, losing its territory on a daily basis. So what kind of standard should be absolute for caging monkeys—how big, how tall, how clean?

There's also the matter of entertaining the caged monkeys. The act also requires that laboratories "enrich" the lives of the captive primates. Again, USDA has allowed flexible performance standards. Animal advocates, fearing that researchers will be grudging about providing enrichment, would like set rules here, too. They want to guarantee that every monkey has something to while away the hours. The devices considered are many: perches, climbing structures, puzzle feeders, kong toys (a rigid rubber tube with a hole in the middle that can be packed with raisins or nuts frozen in ice, which the monkey has to work to remove). One NIH study looked at the merits of putting Astroturf in cages, sprinkling nuts into it, and allowing the monkeys to forage. "My fear is that if we dictate enrichment, we'll end up with regulations that say, in essence, you have to have three kong toys and two perches per animal," DeHaven says. "If facilities meet the minimum, it may do nothing for psychological well-being, nothing for certain species." His voice grows weary with experience: "And we'll have facilities that will meet the minimum and nothing else."

Will toys and perches make a monkey happy? It's a question that alarms Bernstein because he fears—rightly—that many people equate "psychological well-being" with happiness. "I would never write a law saying, 'provide for psychological well-being,' " Bernstein says. "I'm a psychologist and I don't know what that means. When I took the chairmanship of the committee, I wrote to all the people on it and I said, 'What does it mean when I say, "How are you?," and you tell me that you're fine?' Does that mean you're in the best possible physical health you could ever be in?" He shakes his head. "What you mean by that is, 'I'm able to function at the level I expect of myself. I can do all the things I want. I may have a hangnail, a little cold, but I'm fine really.'

"Now, you might look at me and say, you're not fine, you're feverish, your color is bad. You can disagree, fair enough. Can you do that for an animal? An animal always tells you that it's fine. Most animals don't let on that they're sick. You can look for signs that they aren't functioning normally. But I cannot for the life of me come up with any scale to measure happiness in monkeys."

There's also the argument that everyone is making this too hard.

If you can't judge a mentally healthy monkey, it's fairly easy to pick out a stressed one. The one consistent pattern is that monkeys caged alone often show extreme distress, recalling the work of Harlow. It doesn't take much to conclude that a rhesus macaque who rocks ceaselessly, continually pokes himself in the eye, huddles, rips out hair, bites open his arms, is not a normal monkey. You could probably safely say it isn't a happy monkey. The UC-Davis group took 12 rhesus macaques, most of whom had been singly housed for at least three years. They then compared their behavior with animals who had been kept in groups or at least with one companion, often called pair-housing. The isolated monkeys paced twice as much as group-housed monkeys, huddled in corners ten times more often. Self-abusive behavior occurred only in the singly-housed monkeys, never in the pair-housed ones.

On this point, there is little ambiguity. DeHaven forthrightly opposes locking monkeys in sterile boxes. The animal welfare regulations demand this much: If an institution isn't planning to pair or group-house a monkey, then it has to explain why in writing, and the explanation must be accepted by USDA inspectors. Otherwise, isolation-housing is not allowed.

If there was a great conspiracy by the research community to maintain monkeys in isolation, then Irwin Bernstein might not be chairing the panel. He is adamant on the same point. "The almighty dollar does not rule everything," Bernstein snaps. "A caged macaque is not a normal macaque. And we are not talking about building the Taj Mahal here. We're talking about putting cages together and connecting them. Some percentage of animals put into separated cages are going to turn into self-mutilators. That's a given. How many animals do we let do that before we say it isn't acceptable?

"The fact is, primates are social," he continues. "It's not just that people are social, it's pervasive through the primate order. That's how they cope, with social support. You can give them all the toys in the world, and it's not going be a substitute for companionship. And biomedical scientists who say, 'Oh, but the monkeys can see and hear each other,' are ignoring the fact that a monkey's primary social contact is physical contact. They need to touch." That was one of Harlow's great findings, and the years of research that have followed have only emphasized it, not diffused it. The need to touch is so strong among some monkeys that a study of stumptail macaques found that if they even had a small window in their cage, through which they could reach through and stroke each other's fur, they would stop ripping themselves apart.

Perhaps it is justice, then, that one of the strongest voices for social housing of monkeys comes from the University of Wisconsin, where Harlow worked. Slowly, primate centers are coming to believe in "pair-housing" of monkeys, not just of new laboratory animals but of animals that have been kept in isolation for years. It's a practical solution, in part. Pair-housing allows primate centers to rejigger existing cages rather than ripping them out for enormous social structures. It is one of the answers to the argument that psychological well-being of primates is too expensive.

If there is a "father" of this movement, it is Viktor Reinhardt, at the Wisconsin Regional Primate Research Center in Madison. He is a soft-voiced veterinarian from Germany, a man who believes in civilization, who keeps an electric kettle in his desk, so that he can brew tea in the afternoon and serve it to guests. Reinhardt was not Harlow-trained. He had never worked with monkeys before he came to Wisconsin in 1984.

His first job was in Kenya, studying free-running herds of beef cattle. He was hired to look into whether cattle were more productive—more babies, healthier ones—in carefully managed pens or out on the wild. To what Reinhardt describes as his own amazement, the wild-running cattle were far more prolific. He finally concluded that it was the stress, of being separated, that reduced their ability to breed. After six years in Africa, he moved to Canada to study musk oxen and buffalo. He discovered that, while the oxen were interesting, North America felt like a home to him. After his Canadian visa expired, he went job hunting. The former director of the Wisconsin primate center, Robert Goy, hired him as a veterinarian, interested in a man with a different viewpoint.

From the beginning, Reinhardt hated the single-housing of monkeys. "I was horrified," he recalls. "I thought, this shouldn't be so." He would look at huddled monkeys and he would think to himself, this isn't really a rhesus macaque anymore. It looks like one, but it doesn't act like one. Why not, he argues, just rename it for what it really is, an experimental animal, a laboratory caricature of the real thing. Change its official name from *Macaca mulatta* (the scientific name for rhesus macaque) to *Macaca experimentalis*, suggests Reinhardt sarcastically.

He believes that damaging the animal is bad science. When you alter the animal, you are altering your results in ways that you can never fully understand. Reinhardt supports animal research because, honestly, he doesn't see an alternative. "I'm part of society," he says. "If I could rely on drugs that weren't tested on animals, I would. But we aren't there—that's not a place I can go. If I said I was against animal research, I couldn't take Advil for pain. I'm not in a position where I can live without the medical profession."

He holds the profession to a high standard. He is uncompromising that animals in research settings should be acknowledged as the complex creatures they are: "The term I dislike most, perhaps, is that we 'use' an animal. An animal is not an object. You 'use' a chair maybe. But we are not talking about furniture. We are talking about living creatures, deserving of respect."

The Wisconsin center houses about 1,000 rhesus macaques and a scattering of other species, close to 100 stumptail macaques, 70 marmosets, and about a dozen capuchins. Reinhardt was particularly interested in breaking the barriers between the rhesus, who seemed most distressed by isolation. His co-workers did their best to discourage him. They argued that rhesus were savage little animals, destructive, unsafe to put together in a small space. Reinhardt was warned that the monkeys would tear each other apart and have a good go at him as well.

It was one of those times when Reinhardt believes that being very naive can serve you well. "Having never worked with primates, I knew the literature—rhesus were supposed to be so aggressive—but I had no firsthand experience. I thought I had to find out for myself." On the other hand, he didn't exactly want to bring his employer's animals together and have them rip each other apart. He finally decided that he would begin in the most unthreatening way possible, by pairing an adult monkey with a baby, a cute little baby.

He carefully chose young monkeys, between 18 months and 2 years old, old enough to be weaned, young enough to be outgoing and affectionate in their behavior. Then he put them in cages with adult male rhesus, known best for their surly and unfriendly behavior. To everyone's surprise, including his own, the big males were patsies. They were cuddling and playing with the youngsters within days.

The next problem for Reinhardt was that he was running out of young monkeys, and he'd barely begun to move the population of monkeys together. So, he fell back on basic analogy to adult humans. Reinhardt reasoned it out something like this: If one person enters the home of another, expecting to get along, he doesn't just barge right in. People call, ring the bell, knock on the door, greet each other, establishing a comfortable relationship, or a least a liveable one. If someone crashed right through the door, it would definitely signal hostile intentions. The same should be true with monkeys. You couldn't just throw a strange monkey into another's cage and expect them to get along. A human being would defend its home and self in that situation. So would a monkey. So Reinhardt devised what he considered the macaque equivalent of doorbells and introductions.

He put two animals in a double cage, divided by a see-through partition. "They are so smart, they can establish a relationship within minutes," Reinhardt says. He would wait until the macaques had clearly settled into a dominant-submissive relationship: one posturing, the other bowing back and stretching his mouth into the characteristic fear grimace of macaques. If that didn't happen, if they simply continued to threaten each other, he would try another mix.

Even when the right balance was there, he would not simply lift the partition. He reasoned that the animals might still have territorial feelings about their cages, a "this is MY home" kind of reaction. So he would move both animals into a new cage, where everything was strange to them, except the other monkey. He tried it first with females, then with males, then he set himself to pair-house as much as he could throughout the center. More than 90 percent of the monkeys at the Wisconsin primate center are now pair-housed, up from none in 1984. Sometimes a relationship falters, sometimes an apparently happily settled couple starts getting on each others' nerves. If an injury results, instant divorce. But that too, Reinhardt argues, is not so different from humans.

"Ever known of two people who moved in together and had the relationship fall apart," he asks, half-smiling, eyebrows raised. "I think this is just like a human relationship. You can be in love with someone, live with them day in and day out, and the relationship can still not work. Why should monkeys be more like angels than we are?" If he separates a pair, he tries again with other animals. Some of the rhesus macaques at Wisconsin have now lived as "couples" for seven years.

Beyond the issue of pair-housing, Reinhardt illustrates another shift, not so much in attitude but in power. Like Jeff Roberts, in California, Reinhardt is a veterinarian. Like Roberts, too, he came to a job because of an interest in the animals themselves. "It's the animals, learning about them, that makes the job so fascinating," he says. "They're the salt in the soup." Several decades ago, lab vets were considered scientific support staff only, there to do the bidding of the researchers. Now Jeff Roberts can shut scientists out of op-

erating rooms and Viktor Reinhardt can spearhead an effective crusade to bring all his facility's monkeys together. The changing
times, the animal welfare movement, the new law—all of these have
given veterinarians power. They've used it.

DeHaven is a realist about the law. He likens his small staff of
inspectors to a state highway patrol force. State troopers usually
catch the chronic speeders. USDA inspectors will pick up on the regular abusers. They are unlikely to find the occasional slipup. When
they do, it is usually through a whistleblower on the faculty or because a locally vigilant animal advocacy group has called in a problem. The USDA inspectors are vilified by both sides of this issue.
But frankly, DeHaven believes he needs both, needs cooperation
from the research community and could not function without animal advocates.

"There's no question that the Animal Welfare Act exists because
of animal activist groups, and there's no question that they serve a
useful function," he says. "We rely heavily on the public and animal
protection groups to notify us of problems in facilities. And they do
it. That's not to say that side isn't without its bad apples too. There
are groups that have their own agenda and don't let the truth get in
the way. But moderate groups assist us in doing our job and are, in
fact, responsible for our existence."

None of them may be entirely satisfied, but all parties—activists,
researchers, newly empowered vets, and inspectors—have been affected by the 1985 changes in federal law. Over nearly a decade,
they've made adjustments—especially at the laboratories—in attitude as well as in practice. For another example of the change the
regulations have wrought, and how people are already outdistancing
the rules, consider the matter of restraining devices.

Suppose you want to immobilize an animal without drugging
him. The simplest way is to lock him down. The most commonly
used device is a "chair," essentially a vice made of plexiglass and
metal which locks under a monkey's chin and around his arms, holding him in place. They go by different names: restraining devices,
stereotaxic devices if they grip the head in particular. Stuart Zola-
Morgan uses a stereotaxic device when he does brain surgery on
monkeys in San Diego.

Chairs used to be far more widely used. In the 1960s, for instance,
researchers wanted to study helplessness. They strapped monkeys in
restraining chairs and jolted them with unexpected electric shocks,
measuring their stress level. Today, the chairs are used mostly for

neuroscience work, particularly in studying the brain. Some researchers, for instance, look directly at the living brain by drilling a hole through a monkey's skull and cementing a capped tube into it. They can then lower a hollow needle through the tube and look through it, using a microscope to magnify, seeing down to the individual neurons firing messages in the brain. To do that, a researcher wants the monkey awake—and still. So, in the scientific jargon, they "chair" him. The device clearly protects against brain damage as well as protecting the results.

The research results can be remarkable; helping to define the workings of the brain in startling detail. The animals are not in pain; any part of the experiment requiring a surgical cut is done, first, under anesthesia. Rhesus macaques can become so accustomed to chair work that they can be trained to walk directly to the device and sit in it, calmly accepting food treats for their participation.

Still, the image is ugly. A small animal strapped down while a researcher pokes around in its brain. Most people assume, too, even if the animal appears calm, being chaired is stressful. For those reasons, the USDA included restraining devices in its regulations on psychological well-being. A researcher who wants to use a "chair" must now provide a written justification for doing so. The law also requires that a chaired animal get rest periods.

That satisfies the law and USDA. The rule is what DeHaven would call a good minimum standard. At Southwest Foundation for Biomedical Research, in San Antonio, Texas, it isn't good enough for Dee Carey. Carey is a veterinarian and a researcher. He is deeply involved in studies that use baboons to find out if you can protect an unborn child from disease by vaccinating the mother. And like Roberts and Reinhardt, he sees himself also as an advocate for animals.

Southwest specializes in baboons. It is one of those facilities that enjoys an easy climate that monkeys can enjoy as well. The housing takes advantage of the hot Texas weather; most of the animals are kept in outdoor cages or corrals, all big enough to provide for a family of large primates. Still, some of the animals are kept isolated. There are chimpanzees used in AIDS research, housed in individual huts. There are young baboons abandoned by their mothers in the nursery. There are big animals in surgical experiments. Some isolation is a necessity, or an unavoidable circumstance stemming from the research, but Carey doesn't defend it. "Isolation is the cruelest punishment that we devise for people," Carey says. "And these are social beings, too." Some of that attitude is basic practicality. The scientists at Southwest do heart research; they study controlled stress. They don't want to introduce unnecessary stress on the animals. And some of it is what Carey believes is the growing awareness of the new generation of lab-animal doctors.

"There's no question, it's an evolving science," he says. "Part of the problem is that the mentality has been that we, as vets, are doing housekeeping. That's not true. If lab-animal doctors think all we have to do is put an animal in a cage, wash it, feed it, and make it available for scientists, then in my mind, they're not living up to their responsibility." He worries that people get so hung up on meeting standards, guidelines, and regulations that they forget to think, to exercise basic common sense. "The Silver Spring incident is probably an excellent example of how not to do science. Taub— the way he cared for his animals—may have met all the standards. But somewhere along the line, someone with authority has to step in and say, this is enough. That there may be projects like Silver Spring, they may be valid scientific projects, but that this is enough, the animal is suffering too much."

For Carey, the practice of chairing a monkey steps over that line. He doesn't like electroshock studies either. He once stopped an experiment, fully federally funded, because he hated what it did to the animals. "You could look at them back in their cages and they were miserable," he recalls. "When the grant was approved for renewal and I just said no. They said, 'We're going over your head,' and they did. And my boss backed me up. It's a matter of common sense, when you can see an animal suffering."

Chairing is not permitted at Southwest. When an animal has to be restrained, scientists use a harness that wraps around the midsection. An attached cord restricts movement but doesn't paralyze the animal in place. "That's only my opinion, that [chairing] is stepping over the line," Carey says. "Chairing is still permitted in a lot of facilities. So if I went to another lab, as a responsible vet, I would have to seriously consider, whether they used chairs and whether they would eliminate it."

Animal advocacy from within the system is never easy. Roberts sighs over it. Research scientists, he says, are not necessarily anti-animal. Many of them have also driven reform. The field of psychology—so reviled by animal activists—has produced people like Irwin Bernstein, Duane Rumbaugh, and Roger Fouts, all with a sense of mission about better care for animals. The problem, Roberts says, is that some scientists are so focused on their results that they develop a kind of tunnel vision. "They're very driven. And when you are seeing results, you're not focused on what your experiment will do, long-term, to the animal, if it's going to have a permanent effect. Most of the time, when you pull people aside and explain, then they work with you. Sometimes you get people, who say, 'Well, I don't care about that.' Those are the people who make you angry and those people shouldn't be working with animals."

Like Bernstein, he believes that people who work with monkeys must learn to meet them on monkey ground.

"At the center here, we assign our technicians to keep an eye out for monkeys with abnormal behaviors, like self-biting, mouthing their hands, so that we can target those animals for special attention," he says. "We had one tech who then walked right up to each cage and stared at the monkeys. That's a threat to them. So they'd scream and bang on their cages. She reported them as hyperaggressive. And they weren't. They were just responding to the visual cues from her."

Roberts continues: "The thing about working with primates is that, 90 percent of the time, if they understand what you want, they'll do it. You have problems when you give them mixed signals. Say you're trying to get a monkey into a transfer box, out of its home cage. And it's not doing it and you get exasperated and start yelling. Well, the animal may be so confused by the threatening signal it gets from your loud voice, it doesn't realize it can end it by getting in the box. It's stressful for the animal anyway, and you need to give him positive reinforcement. Think about what encourages *you* to cooperate."

Like Wisconsin, the California center is moving as close to total social housing as it can get. Under the new plan, Roberts says, every monkey, barring those in infectious disease experiments, would spend at least part of each day with another animal. His staff has also experimented with many toys and puzzles—the field cages are filled with swings, perches, barrels. And of course, rhesus macaques will make anything into a toy. They unscrew the bolts from the cage; steal wrenches from maintenance workers. They have destroyed so many extension cords that the center now equips its workers with cordless tools. The kick with toys, he thinks, is not just curiosity, it's control.

Primates—human and monkey—need control over their environment, Roberts says. If the monkeys are allowed things they can manipulate—even a toy—then they are allowed some control. The problem in meeting that need, he says laughing, is that sometimes the monkeys outsmart the scientists.

Roberts recently bought a set of puzzle balls for his monkeys, hollow plastic balls with triangular holes in them. The balls were made of half-spheres that locked together. The idea was to drop monkey treats into the balls, through the holes, and let the monkeys work on getting them out. Roberts wanted to make sure the monkeys couldn't just break the balls apart at the center seam. He and two other staffers spent half an hour themselves trying to open the puzzle balls. They wrenched and pulled and couldn't get them open. So, they took them back to the center, gave them to the monkeys, and within a day, some of the rhesus macaques had already figured out how to pop them open. "There were three of us unable to get these apart," he says, unable to keep a straight face. "And this monkey, that weighed maybe 20 pounds, popped it on the first day he had it."

There are more than 3,500 monkeys at the California primate center. Roberts does not believe he will serve each monkey perfectly. Some will become bored with toys too quickly. Some will be stressed by their relationships with other monkeys. That's not the issue. "There's an obligation to these animals to provide them with the best and most humane care, and that [obligation] existed before there was ever an animal welfare law," he says. "The law has speeded things up, but it hasn't changed the obligation."

At the Chicago zoo there were 150 primates. It was easier to recognize them as individuals. Roberts admits that it's easy, too easy, to let the monkeys blur into an assembly line at a big primate center. Still people who like animals should be at animal research institutions, Roberts argues. The alternative is so much worse. There's one other difference, though, between working in a zoo and a primate research center. If you work at a zoo, people believe it when you say you like animals. When you work at a research center, where monkeys are surgically altered, infected with lethal diseases, dissected—it's a harder sell. Sometimes an impossible one.

"Bottom line, we take healthy monkeys and make them sick, you can't get away from that," Roberts says. He used to hesitate when he introduced himself. He'd go to a party, and say, "I'm a lab-animal vet," and he recalls, people would say, "Oh, well." With the implication, "I know what you do, Dr. Frankenstein." For a while, he started ducking it. If he went to a party, he'd just mumble about being a veterinarian. But not anymore. "I say, 'I'm a lab-animal vet and I work with primates, and we work with treatment of primate disease, and we look for treatment of human disease and we do behavioral studies and we try to understand more about the world.' And if they can't deal with that, that's their problem."

His voice softens as he goes on: "I wish I could say, yeah, we're not going to need primates in research. We're going to come up with all the alternatives and models to eliminate them. But I don't believe that."

NINE

Not a Nice Death

Roy HENRICKSON KNEW all about watching monkeys die.
They died all the time when he first came to the California Regional Primate Research Center. It was almost routine. Monkeys
bled to death from fight injuries. Monkeys wasted away from incur-
able diarrheas. Before Henrickson arrived, almost 500 monkeys had
been killed by Simian Hemorrhagic Fever, a racing infection that
ripped blood cells apart, causing the animals to drown in their own
fluid. The center's death rate was 18 to 25 percent in the early
1970s. Losses were so dismal that the federal government threatened
to close the place down.

Henrickson came on as chief vet in 1972. His predecessor had
quit, burned out. That might have worried another veterinarian. It
struck Roy Henrickson as a challenge. He was Hawaiian-born, a big
man with blue eyes, dark hair, and such a passionate love of travel
that he plastered maps across every available wall of every office. As
a primate vet, he'd traveled to India and Malaysia, to study and
understand the animals in their native homes. Earlier, Henrickson
had lived in the Chilean Andes, working in a Trappist monastery,
caring for the monks' animals. He'd tried private practice; it bored
him, and so did the dogs and cats. He'd considered zoos but thought
they were too political. The primate center, with its chattering monkeys and desperate need of help, intrigued him.

The setup at the center was primitive, Henrickson remembers.
Cages were concrete slabs, rimmed with iron bars. No one thought
about monkeys liking to perch or climb. Old tomato boxes had been
tossed in for shelter. It was seat-of-the-pants veterinary medicine at
first. Henrickson packed his instruments and supplies into a suitcase
and lugged them from cage to cage. He also set up a regular clinic.

He isolated ailing monkeys from the well ones. He left healthy colonies alone, refusing to shift them around. He fussed over diet, reading nutrition labels on monkey chow bags, adding in fresh fruit. Within a few months, the accidental death rate had fallen below 5 percent.

He was still feeling smug about it when the monkeys started dying again.

It was different this time, a wild, lethal slide out of control. It was early fall in 1976, the days still cooking with the heat that lingers past a Central Valley summer. The center had a colony of stumptail macaques then, cousins of the rhesus macaques. The stumptails were sweet animals, Henrickson says, more trusting than rhesus and by far more gentle. In the crisping heat of summer, the baby stumptails would sunburn, their small faces brightening to deep pink. Henrickson would painstakingly smear sunscreen and ointment on them and they would cling, looking up while he rubbed the cream in. "They were beautiful animals," he says. "I really liked them." They were suddenly dying away and whatever it was, it was a terrible death, a cascade of infections, cramming one on top of each other, wearing the little monkeys out.

"I'd been thinking, 'Well, you're pretty good, Roy. Sort of a hero.' The disease had really just fallen off because we'd quit moving the monkeys around so much, mixing them and their infections together. It was the equivalent of closing down the gay baths in San Francisco. And [then] this hit the stumptails. I was pulling my hair out. I thought it must be something in the soil. I was out there with a shovel digging up dirt, sending it off to be tested. And the stumptails, they were just babies and we couldn't save them. I can't tell you how much you mind something like that. It was enormously stressful."

The stumptail epidemic passed without ever being diagnosed. Henrickson, angry and frustrated, saved blood samples and slices of tissue from the lost monkeys, stored them in the center's supercooled freezers, ice solid at 70 degrees below zero Fahrenheit. Then in the early 1980s, monkeys suddenly started dying again. This time it was rhesus macaques. This time, Henrickson was determined not to let the disease, whatever it was, get away from him. He assigned a postdoctoral assistant full time to do nothing but clinical workups on each animal, looking for the common pattern.

When the pattern emerged, it would mean more, much more, than the deaths of a group of monkeys in Northern California. It would weave itself into a much bigger picture, a frightening one. At the New England Regional Primate Research Center, near Boston, a similar epidemic was also destroying macaques. And this time, there was something similar spreading into the human population. A troubling illness was emerging, a disease that caused a lethal collapse of the immune system, crippling the body's ability to fight off infection. The human disease was AIDS, Acquired Immune Deficiency Syndrome.

The New England scientists realized, and then the California researchers, that the monkey disease was almost a mirror of the human one. Like AIDS, it was a virus that eventually crumbled the immune system into unworkable bits. All the infections—in the stumptails, the rhesus, the humans—sprang from the same family of retroviruses. Retroviruses can seem both primitive and sophisticated. They are too basic to contain DNA; they rely on the genetic structure of their host. They can whip the human genes around, though, like a ringmaster at a circus. If a retrovirus wants to make copies of itself in human cells, or monkey cells, then it gets the copies made. The human virus is best known now as Human Immunodeficiency Virus (HIV). The monkey virus, the mirror, is Simian Immunodeficiency Virus (SIV).

When Henrickson and his colleagues in Davis went back and analyzed the frozen blood of the lost stumptails, it was SIV they found locked into the cells. No one is sure where, exactly, HIV moved out of the dark and into the human population. But in macaques, there's no doubt. Their version of the disease began at the hands of humans. SIV in macaques is a disease born of captivity. It was carried by African monkeys who were packed into research centers with Asian macaques. Most probably, researchers think, it was transferred by the casual handling of animals, such as reusing needles. The most clear-cut case occurred at the Tulane Regional Primate Research Center. Scientists there tried to infect rhesus macaques with leprosy by injecting the tissue from sooty mangabeys that carried leprosy bacterium. They didn't realize at the time that mangabeys were silent carriers for SIV. The macaques became much sicker than anyone had expected.

At the California center, the evidence is purely circumstantial: Back in the 1970s, there were sooty mangabeys at the Davis site; their frozen blood samples also reveal the knotted presence of SIV. Nobody had known anything was wrong with them. Again, the mangabeys exhibited no symptoms. Scientists now believe the African monkeys were exposed to the immunodeficiency viruses thousands of years ago; they have learned to live with them. Like humans, the macaques are newcomers in retrovirus county, hopelessly vulnerable.

The monkey model for AIDS, then, was created by mistake. It was there, undiagnosed, before the human disease became suddenly visible. It was there when Roy Henrickson was frantically digging up the dirt on the laboratory grounds. Without the human disease, this would have been considered a mistake in monkey management—the loss of valuable monkeys to a stray virus. But with AIDS at large, things were different. When they realized what they had, researchers regarded it as a gift. What incredible timing. Just as the human disease spiked up, so did the monkey ailment. Exactly when they needed it. They seemed to have the perfect model.

In the beginning, scientists thought they had the virus trapped; some virologists were predicting a vaccine by the mid-1980s. Now, there is doubt a vaccine will be developed before the year 2000, or even until the next century. Perhaps the monkey model for AIDS was oversold and perhaps it never was a perfect model. The story, though, also illustrates the limits of animal models, the tradeoffs of using animals to solve human problems. After the splashy successes—the vaccines for polio, measles, mumps—perhaps we were too sure of conquering all. The AIDS virus, if nothing else, has proved the limits of the tools at hand.

The progress slowed to the point that Henrickson himself, ever restless, left Davis for UC-Berkeley in 1985. He wanted a new challenge. "It was tremendously exciting in the early days of AIDS research," he says. "But after a while, I realized I'd ridden the high." He's still at Berkeley, and it's fair to say that the big, Bay Area university, surrounded by militant animal activists, has provided Henrickson with endless challenge. He plans, though, to take early retirement by 1995, and spend his time consulting and traveling. He already has invitations to a primate colony off the coast of Java and to a research station in Kenya.

The struggle to overcome—or even understand—the AIDS virus has been left to those with patience. Or stubbornness. Or both. Back in Davis, virologist Nick Lerche (pronounced LAIR-key) has been working with the primate virus ten years now. Lerche heads the California primate center's Simian Retrovirus Laboratory. He remains a believer that the basic design of an anti-AIDS vaccine will come in macaques. But when? "I go up and down," Lerche says. "I think we were naive in the beginning. We didn't understand the complexity of the immune system or the biological systems we were working with. I think we'll find it in the monkey model, but it's going to be slow. . . . Originally, everyone wanted to hit the home run and we thought we could. The funding agencies thought so, too. It just hasn't happened. The easy stuff has all been done. Now we're having to step back."

Lerche thinks they don't really even have the model yet. Not the one that will unlock the disease. How can they? They don't fully understand the virus or its family; they can't completely explain the body's response to it. Not the monkey response, not the human response either. After ten years, they're still trying to figure out what the model should be. When pushed, Lerche searches carefully for a word to describe where they are. "Evolving," he chooses finally. "The model is still evolving."

Any animal model in medical research is imperfect. A monkey is not a boy; a cat is not a girl. The animal is being experimented on for ethical reasons, as an acceptable substitute for the human being. If you want to study the effect of drugs on the lungs of infants, you don't slice up the lungs of human babies. You use the lungs of baby macaques instead. Heart transplants were developed in dogs. Vision studies are done in cats, peeling open the brain to study visual nerve development. Tests for poisons, checks for carcinogens, these are routinely done in rats and mice, not in people.

If you can't get the human result, you get as close as you can, edging across the species, beginning in rats, working up to primates. You hope that your animal will be a good "model" of what happens in humans. Sometimes hope isn't enough. The thalidomide tragedy proved that people could take a beating from chemicals that were often shrugged off by rats and rabbits. Sometimes, again, the animal results are chillingly accurate; people being people, sometimes they ignore them.

The potent anti-acne drug Accutane is a derivative of vitamin A. Repeated tests in monkeys showed that vitamin A compounds caused birth defects, sometimes bad ones. When Accutane was introduced by Hoffman-LaRoche in 1982, there were warnings on the labels: no use by pregnant women. Yet, doctors prescribed it anyway and expectant mothers used it despite the warning. By the late 1980s, the U.S. Food and Drug Administration estimated that there were 1,000 children in the country with minor Accutane birth defects and another hundred seriously harmed. Serious meant their ears were set into their jaws, they were perpetually dizzy, the balancing mechanism of the brain improperly formed.

In the case of Accutane, you can make the argument that the animal model wasn't good enough because people didn't pay attention to it. It took the human model, paying a price in children's health. It took human suffering, the very thing animal testing was meant to avert. Only then, people believed.

At some point, and this is particularly true with AIDS research, if you are going to believe in the animal model, you have to also accept that an evolving animal model is not a failed concept. When people compare scientific discipline to religion, it is because of the qualities both share—a faith in the unseen, the fundamental arrogance that comes in believing you are right. It takes both, faith and arrogance as well, to believe that after ten, frustrating years or more you are still going to wring the answers to AIDS out of some sick monkeys. It takes charting process in tiny steps, not in bounds or leaps. A believer argues that maybe we don't have the vaccine answers yet, but parts of them. Researchers have begun stripping apart the virus; they have learned some of its strategies, some of its defenses. You keep believing and adding together those small pieces; sooner or later, they may add up to something, like a solution.

It's a treasured position of research scientists that you have to be patient; you have to allow unfettered curiosity and the occasional aimless ramble of the trained mind on the way to a solution. You have to accept that there will be deadends.

Perhaps the most compelling illustration of an animal model in evolution—illustrating the need for patience—is the erratic history of atherosclerosis work in baboons. The baboon model has a purely serendipitous beginning—at a zoo in the Louisiana river delta.

Henry McGill, chief of research at the Southwest Foundation for Biomedical Research, remembers exactly how baboons became the model of choice. When McGill first came into medicine in the 1950s, neither he nor any other heart researcher thought about using baboons. At Southwest, in San Antonio, the popular model was the heart of a calf; researchers there were cutting hearts out of young cattle and putting them into machines that pumped fluid through the hearts, watching how they worked. There are no cows now at Southwest.

The facility spreads across some 200 startlingly green acres in urban San Antonio, with the bright banners of SeaWorld blowing in the background. Southwest now dedicates itself to primate research. It houses more than 100 macaques, 220 chimpanzees, and close to 3,300 baboons, the largest colony outside Africa.

In 1956, McGill was a pathologist at Louisiana State University's New Orleans division. His office was across the street from Charity Hospital, the city's big, teaching hospital. It so happened that another pathologist at Charity had a fascination with exotic animals. When the residents of New Orleans's Audubon Park Zoo died, the researcher would rush over, cart their bodies into the hospital's autopsy room, and cut them open. Tigers, zebras, lions, alligators, and monkeys had all been autopsied in the morgue of Charity Hospital. "Those were the good old days," says McGill, half laughing. One summer morning, he got a call from the pathologist. The man had been exploring the body of an old female baboon. He knew that McGill was interested in atherosclerosis, the disease which builds up fatty deposits in the walls of the heart. "I've got a heart you might like to see," he said.

"So, I trotted across the street," McGill said. "He had dissected the aorta out and it did have some fat deposits and scars, very similar to mild human atherosclerosis. It was very interesting because this animal had lived in the zoo for many years; it had been captive-born, I think, and had never eaten anything but monkey chow, peanuts, and popcorn. Yet, there were the deposits. So, I picked the aorta up, put it on a piece of cardboard, and trotted back across." The chief of McGill's department and Nick Wurtheson, a scientist from Southwest, were meeting at the time. "I can still remember this like it was a photograph," McGill says. "I remember going in, holding out the heart, and saying 'Look at this!' And they got very excited. And I was pleased because I was excited, too."

It has taken more than 30 years and several career moves—eventually bringing McGill himself to Southwest—but by tracking natural atherosclerosis in the animals, the Texas scientists have helped pick apart the genetics of heart disease. They've discovered that fatty streaks in the heart seem a process of nature, a natural by-product of the way the body cycles fats through the blood. "But what separates the sheep from the goats," McGill says, "is what happens to those fatty streaks." In many animals and people, they remain harmless little dots of fat. But in others, suddenly they turn into fibrous, destructive scars, clogging the passages of the heart. In humans and baboons, that change toward destruction begins, suddenly, in young adulthood. And whatever drives the shift seems to be largely programmed, a consequence not just of diet and exercise, but of individual genetics.

They know that because McGill and his colleagues have been carefully breeding baboons for several decades. They've interbred animals with a tendency to develop heart disease; they've mated animals with hearts that appear unusually resistant to heart damage. They're on the fourth generation of those select baboons. They've discovered families that, no matter what they are fed, never develop high blood cholesterol, never go onto heart disease. And they've discovered baboon families that develop soaring cholesterol levels at the drop of a cheese slice.

They have been able to identify one of the proteins that seems to induce high cholesterol. They've already applied for a patent on it, hoping to produce a drug to counter the effect in humans. But the real breakthrough will come, McGill says, if they ever can figure out how the low cholesterol baboons buffer themselves. He has no doubt that, if they do figure that out, they'll be able to transfer the knowledge to humans.

The genes that control the heart are about 93 percent the same in humans as in baboons, McGill says. The proteins that make up the heart muscle are almost 98 percent alike. "It's been a long process and we've had trouble getting funded sometimes," he says. "But we're just entering the payoff period now."

Some of the payoffs have been unexpected. Because of maintaining a long-term breeding colony, Southwest has some old baboons. When they die of old age, McGill routinely performs a necropsy, essentially an animal autopsy. He noticed that with elderly females, the bones were extremely soft, so soft that he could cut them with surgical shears. That fact was at the back of his mind when he served on a national committee on health concerns and sat next to a scientist studying osteoporosis. After comparing notes, McGill began a survey of his aging female baboons and discovered that, like humans, they go through both a spontaneous menopause and, following that, a rapid decline in calcium in the bones and the resulting softening of the bones known as osteoporosis. Like aging women, female baboons develop compression fractures of the spine, fragile bones, even the classic, hunched-over back.

"In most cases, there is no single best animal model for a human condition," McGill says, reflecting on the unusual circumstance that the baboon appears to be the best model for not just atherosclerosis, but also osteoporosis. "In this case, they are just like humans."

Yet if you had evaluated the baboon studies after their first ten years, you might have written them off as a complete waste. You might have complained that baboons had been yanked out of Africa so that baffled scientists could struggle, unsuccessfully, to figure out their hearts. But, even then, McGill knew where he was headed. He just had to figure out the route. McGill was able to do that without public impatience and doubt. He held an advantage that AIDS researchers are denied: He wasn't on center stage, working hurriedly in the middle of a widening epidemic, trying to figure out the virus from hell and struggling to make sense of a model that isn't a perfect fit.

If you accept animals as the ethical substitute for humans—and that can be a big if—then maybe you just prepare to wait out the monkey model of AIDS.

And maybe not. Maybe you wonder why the disease continues to run out of control, even as the federal primate centers spend a third of their $40 million-plus annual budget on AIDS research. Maybe, as animal advocates do, you start thinking ethics isn't even the fundamental issue. Maybe it is that animals are a lousy model for this disease, that they tell us much about how monkeys get sick, and nothing about how retroviruses chew their way through the human immune system. The work may have told scientists a lot about retroviruses in rhesus macaques, or crab-eaters, or pigtail macaques, or even chimpanzees. Has it told them anything about the people dying alone on the wards of urban hospitals?

Activists think not. They see two things going to waste: money and lives. Their attitude can be summed up in the title of a short essay in the quarterly magazine of the *New England Anti-Vivisection Society:* "Experimenters Fiddle While AIDS Rages."

In other words, primate research on AIDS is a vulnerable target. In their early certainty, researchers helped make it one. Virologist Robert Gallo, of the National Cancer Institute, once announced that if he just had 500 chimpanzees, that's all it would take to guarantee a vaccine against AIDS. Obviously, it's not that easy. Chimpanzees now often seem a dubious model for the disease. They are also rare, so their use is politically sensitive. Researchers such as Roger Fouts have publicly denounced putting a dwindling chimpanzee species into further jeopardy in order to make almost no progress against the disease. Chimpanzees in AIDS research perhaps serve best as a haunting example of the biological tradeoffs of being almost human.

From the beginning, researchers thought the perfect model for AIDS would require the human virus—putting it into another animal, such as a monkey. There are actually two variants of the human AIDS virus, christened HIV-1 and HIV-2. The second is still grounded in Africa, slower to move, milder in its course. HIV-1 is the killer and the invader, moving from country to country in a way that reminds people how little political boundaries are worth. When the studies first began, scientists tried to infect rhesus macaques with HIV-1. Nothing. The monkeys blew the virus off as if it didn't exist. Virologists took samples of spleen, lymph node, bone marrow, brain, and plasma from AIDS patients and injected them into other primates—crab-eating macaques, stumptail macaques, capuchins, squirrel monkeys. They got nothing.

The only nonhuman primate that could be induced to respond to HIV-1 was the chimpanzee. Inject the human virus into chimpanzees and they promptly booted up an immune response. You could track the antibodies, swarming against the AIDS virus, in their blood. They didn't get sick, though. At first, scientists thought perhaps it was just that HIV had a long latency in chimpanzees, as it does in humans. Now, there are some 70 chimpanzees plugged into AIDS research programs. Some of them have been HIV-positive for eight years. None of them have shown even a quiver of actual illness. No one understands it. Why don't primates so nearly human become sick like humans do?

It makes researchers themselves uneasy, despite Gallo's earlier call for chimpanzees. "Well, there are always tradeoffs with animal models," Lerche acknowledges. "And with chimpanzees it's that they don't get sick." But if it makes virologists like Lerche unhappy, it makes animal protection groups absolutely furious. A chimpanzee that goes into AIDS research is a tainted animal. It is, after all, injected not with the monkey virus, SIV, but with HIV-1, the immunodeficiency virus most lethal to humans. Although the chimpanzees don't get sick, they still carry the virus. Their blood is still dangerous to people working with them. An HIV-positive chimp is handled in isolation. It is separated from other chimpanzees to minimize the risk of accidental spread of infection. A chimpanzee can live out its lifetime, apparently healthy, yet an AIDS carrier, boxed up alone. The whole crusade against animal AIDS research by SEMA, Inc., in Maryland, the uproar that so engaged Roger Fouts and Jane Goodall, began with chimpanzees sealed away in the name of AIDS research.

The apparent value of chimpanzees is in vaccine testing. Does a vaccine rev up a decent antibody response, powerful enough to knock off an invading virus? You can challenge a chimpanzee with a vaccine, track its antibody levels, inject the actual viruses, and measure, cell by cell, whether the vaccine has offered any protection. Lerche's suggestion is that chimpanzees be used as the end test, the gold standard for a vaccine. When, and if, a vaccine approach is hammered out in macaques, it gets a trial run in chimpanzees.

It's the mainstream attitude and, on this point, there are plenty of people out of the mainstream. The author of the essay that compares AIDS researchers to the Emperor Nero, famed for playing his violin while Rome burned, is one of them. The essay was written by a registered nurse named Betsy Todd. Todd is an instructor of nursing at the College of Mount St. Vincent in Riverdale, New York. She is also a member of Steve Kaufman's group, the Medical Research Modernization Committee. She gained a lot of attention in the animal advocacy movement with her 1991 master's thesis at the Columbia University School of Public Health. Todd gave it the neutral title "Animal Research and AIDS." Under that polite cover, it is a scathing dissection of the failure of animal models to shed light on the black plague of the AIDS virus. The manuscript is unpublished, but photocopies of it have sailed back and forth across the country, from the offices of PETA in Washington, D.C. to the desks of IDA in San Rafael, California.

Todd's basic premise is that animal models are lousy predictors of what happens to humans. Penicillin, that standby human antibiotic, is toxic in guinea pigs and hamsters. Aspirin causes birth defects in rats and mice and poisons cats. She breaks down the thalidomide effects—teratogenic (causing birth defects) in some rabbit breeds and 7 primate species. Not teratogenic in at least 10 rat strains, 15 mouse strains, 11 rabbit breeds, 2 dog breeds, 3 hamster strains, and 8 other species of monkeys. "It should be no surprise then," Todd writes, "that 51 percent of the 198 drugs approved by the Food and Drug Administration, from 1976 to 1985, caused serious postapproval adverse reactions, including permanent disabilities and deaths." The source for those figures is a 1990 General Accounting Office study.

Further, she complained, the science community wants it both ways—total faith in animal models when expedient; total uncertainty when not. Todd gives two examples: In 1982, the British

Ministry of Health refused to approve the birth control pill Depo-Provera, citing evidence that it caused cancer in monkeys. The manufacturer, Upjohn, promptly declared that the animal tests were misleading. In 1989, the pharmaceutical giant Burroughs Wellcome sent out a letter to physicians who treat AIDS patients, explaining that mice and rat studies showed the popular drug AZT to be a weak carcinogen. "It is not known whether the drug causes cancer in humans," the letter added. That same schizophrenic attitude has run through the work on primate models of AIDS. MicroGeneSys, a private biotech company, tested an anti-AIDS vaccine in chimpanzees. The vaccine used a protein from the virus' outer envelope. It failed to get a reliable antibody response from the chimpanzees. But researchers still believed the concept to be good. The company has continued to pursue the vaccine.

Todd doesn't like any of the animal models for AIDS: rats, mice, cats, macaques. The chimpanzee model, however, she judges to be no better than science fiction. The government began its chimpanzee breeding program to assure a supply of the animals. Government officials estimate, though, that the program produces, maybe, 25 chimpanzees a year. Each one costs between $60,000 and $100,000. Only an unusually well-financed scientist can scrape together the money to use as many as half-a-dozen chimpanzees in an experiment. Most researchers are lucky to afford a couple of chimpanzees, and it shows. Optimistic papers have been published based on results from only two animals.

Exasperated statisticians have pointed out that for a study to have reliable numbers, showing that a vaccine fails 10 percent of the time, a minimum of 29 chimpanzees would be required. Two is meaningless. Todd's second point, though, is this: What if you did have 29 chimpanzees, their blood swimming with HIV, their bodies barely touched by viral effects? What can be learned about the disease from an animal that doesn't get sick? The virus is more sluggish in chimpanzees; it doesn't cause as much cell death. In humans, HIV seeps into spinal fluid, the brain, saliva. In chimpanzees, Todd says, the virus has not been found in any of those places. So, you can toss a vaccine in, look for protection, but does it actually mean anything?

If one doubts that scientists themselves are unhappy with the chimpanzee model, one has only to look at the desperate push to infect another animal, any other animal, with HIV-1. When scientists at the Washington Regional Primate Center in Seattle announced that they had found a chimp alternative in 1992, they announced it with pleasure, and it was received that way as well.

The center happened to have a colony of pigtail macaques—big, tree-dwelling cousins of the smaller rhesus monkeys. Pigtails showed a strong response to HIV-2, the lesser human virus. So the Seattle scientists decided to push, to see how they reacted to the killer version. They injected four strains of the HIV-1 into pigtail white blood cells. Even with their hopes up, they were startled by the way the virus roared into the cells, destroying as it went.

The results were so good, the researchers immediately began rechecking them. They took two pigtail macaques and removed white blood cells from each animal. They mixed the cells with a strain of HIV-1 and put the mixture back into the animals. For up to a year, they periodically withdrew blood from the monkeys and put it into healthy cell cultures. The virus immediately flared in the cultures. Cautiously, the team deliberately infected another six pigtail macaques. Not only did those animals show antibody response, but they showed suspicious signs of AIDS-like illness: swollen lymph glands, rashes, diarrhea. It was at that point that the Seattle group happily announced their findings. The researchers suggested that soon the two-chimpanzee model would be replaced by the 20-pigtail-macaque model. All hopes were high that they were right and researchers began putting in their requests for pigtail macaques, leading one importer to describe them as "the hottest monkey on the market."

The Seattle group remains optimistic about the pigtail promise, but they've encountered growing scientific doubt. In larger tests, pigtails have not gotten consistently sick or stayed that way. In some animals, the infection seems to be a passing thing. In many of the monkeys, antibodies against the virus show up and then seem to just start disappearing, as if the virus had made a feint and backed off. "I'm skeptical," admits Lerche. "It doesn't seem to be living up to the potential or hopes of that model. What I'm hearing now is yes, you can infect, but the infection is transient. . . . It doesn't seem any better than the chimp model. Actually, if it's transient, it's not as good as the chimp model."

One of the angriest arguments against AIDS and animal research is that, at this point, with research staggering along, the government would do better to put more money into education and treatment. Take the roughly $10 million a year that the federal primate centers alone spend on AIDS research and put that into hospitals, hospices, preventive education. There may be no drugs and vaccines yet, but if you could persuade people to routinely use condoms, you'd hold the viral spread to a trickle. Todd complains in her paper that officials are slow to fund prevention and health education, arguing that it's difficult to measure success. Why, she asks bitterly, does animal research always get the benefit of the doubt? Why aren't prevention and education given a chance? Against all odds, scientists will hold to an animal model that seems to tell them nothing, taking money that could be used in better ways, even something as simple as encouraging wider use of condoms.

"There are some questions science is simply unable to answer, or at least, areas in which the answer will be slow in coming. This is not a reason to stop trying, but also no reason for continued attempts to ram what appears to be a square scientific peg into a round hole," she writes.

Of course, your patience depends partly on whether you are waiting for the big answer or willing to settle for a small piece of it. One of the studies that Nick Lerche is working with involves a specific question about the virus. As the number of AIDS cases in women rise, so do the number of cases in children, starting as fetuses who get the virus from mothers. About a third of the time, a pregnant woman will infect her child. But how? And why only sometimes, not always?

So, Lerche began to wonder what the virus does to the unborn child. In humans, it's difficult to separate that out. Many of the HIV-positive women are drug users. They are often sick, rarely following a USDA-approved, balanced diet. All of that bears on the baby. Maybe the mother's poor health is what creates the small and sickly infant, often born to AIDS-infected women. To find out, you basically need a sick baby in a healthy mother.

Lerche and his colleagues have created that, using rhesus macaques. At first look, the rhesus would appear to be the wrong model. Its placenta is too different from the human, a curved double-disk shape as opposed to a single disk. The California scientists, though, have made the placenta essentially irrelevant in their study. Using ultrasound, they bring the shadowy outline of a fetus into focus, then inject SIV directly into the tiny monkey, still inside his mother. During the pregnancy, the mothers are coddled, cosseted, well fed. Yet, consistently the babies come out small, their body weight down. Obviously, then, the virus harms the fetus, even as it grows within a healthy mother.

"What the animal model is for, is to answer questions that you cannot do in humans," Lerche says. "This is the first clear indication that the virus, alone, can affect growth in the fetus. How else do you learn that?"

Critics complain that monkeys don't get sick as people do. The virus races through macaques much faster than it does in humans. They fall ill within three months compared to three years or more in humans. Monkeys develop more tumors than humans. They are hit harder by bacterial infections. The viruses themselves, however, are startlingly similar in human and monkey. They burrow into cells in the same way. They strike at the same parts of the immune system. They coil into the genes in the same way, too.

There are those who argue that it's playing word games to separate the human immunodeficiency viruses from the simian ones. Among researchers who believe that is Jonathan Allan, an evolutionary virologist in Texas. Allan works, like McGill, at the Southwest Foundation for Biomedical Research. His specialty is retroviruses in all their devious shapes.

"When you get down to it, there's no such thing as SIV and HIV," Allan says. "It's the same virus. We give them labels, but it's really nonsensical. It's only viruses that inhabit different species for particular reasons. We've seriously talked about calling them all PIV—Primate Immunodeficiency Viruses, because that's what they are."

There's something about viruses that grow south of the equator, something in the tropical cookery, that seems to grow them wicked. The breathtakingly lethal killers—the terrifying hemorrhagic fevers, like Ebola—spring out of those southern rainforests. No one knows exactly how old the AIDS viruses are, when they started boiling up out of the tropics. Some say 10,000 years, some say 50,000. There may be as many as 30 distinct types of SIV harbored in African monkeys. It takes time for a virus to splinter like that. Evolutionary biologists now think that immunodeficiency viruses may have moved into African green monkeys more than 10,000 years ago, perhaps even 15,000 years ago. It's partly the splintered effect that makes them think that. It's also because the African monkeys don't get sick when infected with immunodeficiency viruses. African green monkeys have made their peace with virus. So have sooty mangabeys. Nearly half the sooty mangabeys tested for SIV are both virus-positive and healthy.

Naturally, a virus tries to find a balance with its host. That's a parasite's self-protection. Kill the host and it has killed itself. It's when they are in strange territory—an unaccustomed species—that viruses are so dangerous, running amuck, trying fiercely to protect themselves in an alien environment. Transfer the virus from a sooty mangabey to a macaque, and the Asian monkey burns up like a tree afire, the virus blazing wildly. The transfer wakes a sleepy virus in another way, too. The abrupt change can bring up a scramble of genetic changes, part of the coping mechanism. In mutation, a benign virus can become a malignant one. The AIDS virus brings to that picture its own worst qualities. It's a virus that was somehow designed to mutate like a lunatic—the most wildly unstable virus yet known.

There are two critical acids involved in making genes work. Deoxyribonucleic acid (DNA) carries the elaborate genetic coding for life. Ribonucleic acid (RNA) serves as a messenger, bearing information that helps turn the genes on in proper formation. The messenger calls out the instructions to the DNA, telling it what kind of cells to make, how to organize them. They are a team, a genetic relay team. Within the human cell, they try to catch each other's errors.

Retroviruses have only RNA. They insert into human DNA, making use of the cell machinery. Their messages get passed along as well. When a cell divides, the virus is copied, too. The problem is that its instructions don't seem to include anything that would fix mistakes. The retrovirus orders genetic copies as if it's typing on a computer without a delete key. You can only type on. There is no backspace, no back delete key, no cursor control. In AIDS viruses, the copies are full of typos. That means the infection comes not from one virus, repeated many times, but from hundreds of thousands, each a half-step different from the last. "RNA viruses are notorious for rapid mutations and defective copies," Lerche says. "And of all the RNA viruses, it appears that HIV-1 is the premier maker of mistakes. You don't get infected with a virus that replicates itself religiously. You get a swarm of variants."

Allan describes HIV-1 as about 60 percent genetically identical to the virus that snoozes away in sooty mangabeys. On the other hand, that virus—call it SIV sooty—is also only about 60 percent like the SIVs that inhabit African green monkeys or baboons. This is the splintering off of SIV variants that makes scientists suspect the infection has been around a long time in African monkeys. But stack SIV-sooty up against HIV-2, the lesser of the human AIDS viruses, and you have suddenly handed virologists a puzzle. The sooty virus is nearly identical to HIV-2, a match of about 84 percent. When you take into account the normal tumbling instability of any retrovirus, HIV-2 and SIV-sooty are basically the same.

"When we first started isolating SIV here, it was done in human primary lymphocyte cells," Lerche says. "So we knew it grew very nicely in human cells. That's an alarm or a red flag there. Then subsequently came the realization that HIV-2 and SIV were very, very closely related. You really can't tell them apart." That kinship supports the idea that AIDS began in Africa in nonhuman primates, and spread to the human ones.

No one knows the path by which immunodeficiency viruses might have moved from monkeys to humans. One popular theory points out that the development of Africa changed in the twentieth century, from locally isolated villages to spreading cities and continent-wide travel. Perhaps, accidentally, the new generation of travelers carried the virus with them. Perhaps it began with a few infected villagers in sub-Saharan Africa. In those isolated villages, the virus could have been spread from monkeys to people fairly easily. Sooty mangabeys, for instance, are part of the food supply.

"There's been a lot of conjecture and sometimes it's been very damaging to people in Africa," says Jonathan Allan at Southwest. "The only thing I can say is that, here I am in Texas. And in Texas, people go out and shoot deer. They dress them down, use their skin, make all kinds of stuff like deer sausage. And what if instead of hunting deer, they were hunting monkeys with SIV? What if deer had the virus that caused AIDS in humans? Then we'd have AIDS here from deer hunters."

The curious thing, though, is that while such theories exist for HIV-2, the story on HIV-1 remains a mystery. Allan says that genetically the closest primate virus to the human killer is SIV-cpz, a variant of the virus found in African chimpanzees. But it is only found in a few of them, and only recently. The chimpanzee version has not been detected in any chimpanzees in American laboratories. Equally mysterious, if the chimpanzee virus is so like the human one, why doesn't the human one make chimpanzees sick? In all this slip-slide of genetic variation, where's the region that makes HIV-1 a serial killer and HIV-2 a small-time mugger? "If I knew that I wouldn't be talking to you," jokes Lerche. "I'd be preparing my acceptance speech in Stockholm."

Here are SIV and HIV-2, almost identical. Give SIV to crab-eating macaques, and they not only get infected, they get sick. Inject them with HIV-2, and they show antibody response but no illness. The reverse seems true for people. They become ill with HIV-2 but not, yet, with SIV. Two American lab workers are known to be antibody positive for SIV-sooty, infected by accidental needle sticks. For several years now, their doctors have monitored them without seeing a trace of illness. Why, why? This ever-evolving animal model raises questions faster than it can answer them.

It is against that background that you have to consider the hunt for a vaccine against AIDS. When, back in 1984, researchers were predicting a vaccine within two years, they were thinking about polio. Once researchers realized that they could infect macaques with polio, the vaccine was in the pipeline. Polio viruses tend to make pretty good copies of themselves, one close likeness after another. If you think of the body's immune system as a police force, then a vaccine can be seen as something like a wanted poster. With polio, scientists introduced into the body either a dead virus or a partially disabled one, with a known face. The vaccine told the immune system what to look out for, so that it was prepared to nab the invading virus if it showed up, even years or decades later. But what if the invading virus changes so fast that it quickly ceases to resemble the face on the poster? By the time you could develop an AIDS vaccine, it might be useless. The inoculated body's immune system still wouldn't be able to recognize what the invader had become.

Given time, left alone in a species, a virus will move toward becoming less pathogenic, more stable, as it clearly has in the mangabeys. If you left it alone in humans, scientists know it would eventually do the same. But in how much time? When in that 10,000-plus years of life in the African monkeys, did the virus relax into co-dependency? "Listen," Allan says firmly. "People are dying of this disease and that's the fact you have to focus on. So, what I'm telling you is conjecture and very dangerous conjecture. But if you continually pressure the virus, you run the risk of changing its evolutionary path. You keep feeding drugs in, you select for drug resistance [which is happening] and you may have forever altered the path the virus will take. It doesn't go in reverse. So when you push it down this path, it doesn't retrace. It's an interesting idea, I think, that we may be preventing, in some ways, the ability of the virus to become nonpathogenic. You keep jacking it with something, and it keeps responding."

There are two distinct HIVs now. HIV-1, the serial killer, and HIV-2, which seems to move more slowly into the immune system and cause lingering chronic infections, rather than deadly ones. By the time virologists develop a vaccine against the two, will there then be HIV-3, pushed into existence by our frantic efforts to control the disease? The virus keeps careening off into unexpected regions, reminding people again and again that it is not a tidy little self-contained package like the polio virus. It needs the DNA of the host. It winds itself into the cell machinery. In a way, it becomes part of the cell. How then do you design a vaccine for a virus that has laced itself into your own cells? If you target the virus, do you kill the host as a side effect?

At the New England Regional Primate Research Center, for instance, scientists have developed a nonpathogenic strain of SIV by deleting one gene from the virus. Using that stripped-down virus, they've successfully vaccinated a number of macaques. But even the New England scientists have made no claim that this would be a good overall vaccine approach for humans. "That's one of the questions. Will we have a vaccine in the next five years, and if we do, will it be one that's used?" Lerche says. "There's still a lot of basic questions about the New England vaccine. How protective is it, what's the mechanism? A lot of vaccines work in some animals and not in others.

"The other point is, even if it's mutant or attenuated, it's still live virus. It still replicates. And retroviruses are notorious. They integrate into the cell genome. Just having the virus genome in your cells may turn on other genes." What Lerche fears is that a vaccine with live HIV, even supposedly altered to be harmless, still contains a virus that can play with human genes. Left in the body, the altered virus might mutate back into something wicked. Or even in the so-called benign form, it might play a little with the human genes. Perhaps, tucked in the body's cells, it might accidentally turn on a genetic switch. By mistake, it could turn on one of the oncogenes, one of the genetic mechanisms that build cancer. When oncogenes flick on, they order abnormal cell growth, turning healthy cells into malignant ones.

"So you really have to look hard at giving a live virus to somebody. I think this kind of work is good to do. The animal models reveal what's important. And I'm not saying the research on the

New England vaccine isn't good. It is. It's fascinating. But I don't see a live retrovirus vaccine for people." Given the risks, the obvious question arises. Would Lerche, himself, accept a live retrovirus vaccine if that became the first AIDS vaccine on the market? "Myself, personally, no."

It seems—often—that what the monkey model has done best is prove what won't work. Still, Lerche, Allan, Roy Henrickson, many in the primate research community believe that someday it will show what *does* work. They believe in the evolving model, despite the frustrations. Someday, as Henry McGill has done with his patient crafting of the baboon model of atherosclerosis, they believe the monkey model will help them decipher the AIDS riddle.

There is, undoubtedly, an element of faith in that. And certainly, you can argue many things about the primate model. You can deplore its failure to yet end the disease, the dismal track record in vaccine and treatment development. You can also hail the insights, often startling, it has provided into infectious disease and the body's immune defenses. But, there's still no end to it, either to the arguments or to the disease itself. Roy Henrickson may know better now than to dig in the dirt, trying to save his pigtail macaques, but the monkeys would still die.

So, there's something else that drives the model, keeps it alive. Curiously, it may be death itself. If you see a monkey sick with SIV, it seems hauntingly like a person dying of AIDS, fading away. The sick monkeys are so like the sick humans, scientists cannot help but believe that somehow, if they can solve the puzzle of the monkey virus, the barriers to the human puzzle will fall as well. People and macaques *are* alike in this reign of the retroviruses. They both die, each beaten down by a disabled immune system, a body opened up to a relentless attack of opportunistic infections. The patients may look physically different, the disease may move at varying speeds, but the ending is the same. No one wants to quit because the virus still wins. "When the monkeys get the disease," Henrickson says, "they die like people with AIDS do. It's not a nice death."

TEN

Just Another Jerk Scientist

THE EPIDEMIC STARTS so quietly that it almost goes unnoticed. Just a few people in the beginning: a man in Baltimore who slowly loses the ability to control his hands, a woman in California whose face slowly becomes numb and stiff, a Missouri child stumbling when he walks.

The disease comes on gradually. The victim slips unsuspecting from vague achiness into a loss of muscle control. At first, they seem unconnected, that scattering of people with the strange, crippling, tissue-wasting disease. Then there are more people, and more, until they form an ominous thread, wrapping back and forth, crisscrossing the country. And now people are dying, and the CDC then steps in to try to figure out where in the world this strange and terrible disease came from.

Right now, the scenario comes from the mind of Jonathan Allan, evolutionary virologist, sitting in his desk chair, head tipped back a little in thought, blue eyes narrowed in concentration. "Does it remind you of AIDS a little?" asks Allan as he finishes spinning out his dark vision of the future. He looks up suddenly and directly, unsmiling, a thin, intense, fair-haired man.

His office is bright-white, sterile-scrubbed. Outside though, at the Southwest Foundation for Biomedical Research, baboons fill cage after cage, acre after acre. It's a striking contrast: the quiet, glass-and-metal framed laboratories linked by polished floors; the dusty, noisy troops of African monkeys scrambling across rock piles in the dry Texas sun. When Allan weaves his little nightmare, he's thinking of both—of wild monkeys and human medicine. He believes that sometimes they may be a good combination and sometimes a lethal one. Sometimes a monkey is a dangerous being to bring into a sterile lab or a super-clean operating room.

Allan began thinking about his epidemic in 1992. That was the year Pittsburgh surgeon Thomas Starzl used a baboon from Southwest—one of those very animals that clambers on the rocks outside Allan's office—and transplanted its liver into that of a dying man. The question, the big one for Allan, is what else gets transferred with the liver? Does the patient also receive some exotic viruses, imbedded in the monkey tissue? If the new infection does grow in the human liver-recipient, does he pass it along to his wife, a friend, a stranger in the grocery store?

Since drugs and surgery have made organ transplants possible, surgeons have been frustrated, to the point of near fury, by a lack of donors. Nearly a third of the people needing an organ transplant die, waiting. To have it all, the ability and the knowledge, to be stopped just by lack of material, has seemed unbearable.

Transplant surgeons have tried anything and everything that might work. They've used mechanical hearts, plastic pieces of hearts, pig hearts. But they keep coming back to primate hearts and they keep coming back to baboons. If chimpanzees weren't so vanishingly rare, they'd be a better choice. The animals—and therefore their hearts—are bigger; the genes are more closely matched. Baboons are the second choice, but they have a lot going for them, too. There's a 93 percent match in genetics between a human and baboon heart, an almost perfect protein match. And baboons are easier to kill for less tangible reasons, for the very reason that they seem less a mirror of people than do chimps. Heart transplant pioneer Christian Barnaard experimented with chimp heart transplants in the 1970s, but he stopped after two. He told newspaper reporters that it began to make him feel too guilty.

For Starzl, the baboon liver was a straightforward choice. He wanted to save a human life. The man had a diseased and almost useless liver and no donor options. Sometime in the future, researchers think that biotechnology will allow them to grow transplantable organs in farm animals; human-compatible livers in pigs, constructed out of the proper proteins. That's still science fiction, though. The indefinite, visionary future offered no help to a dying man in 1992. The baboon liver was there for the picking. And if this patient died—which he did some two months later—Starzl and his colleagues reasoned, they would learn from the surgery.

It was a clear choice as well to Robert Winters, another liver disease patient at the Pittsburgh center. Shortly after the baboon transplant, about 15 pickets were standing in front of the hospital, chanting "Animals are not spare parts," as the 34-year-old Winters started up the steps, on his way to another round of hepatitis treatment. The hepatitis virus is a relentless destroyer of liver tissue. "I didn't ask for this," Winters shouted at the protestors. "But I've got it, and I'm fighting it just like that guy lying up there. You don't have a right to be standing there." The chanting stopped as Winters continued. "Do you know what it's like to have liver disease? Do you?"

The picketers stayed, though, after he stomped away into the hospital. They returned, more angry, when the Pittsburgh group released the news that the transplant patient had AIDS. In a way it made sense. The man agreed to a risky experimental procedure because he was incurably ill anyway. Whether the AIDS made any difference in the transplant, it's not clear. After all, physicians wreck the immune system of transplant recipients anyway, suppressing it to keep the body from rejecting the new organ. The shadowy presence of HIV did, though, emphasize the value system involved— that even a man with a terminal illness was worth more than a baboon. That the baboon was basically a supplier of parts, bred at Southwest like farmers grow tomatoes for salad.

In the shadow of that ethical uproar, Allan's more pragmatic worries were barely heard. Yet his point was not so different. He also thought that this was far more than a simple liver transplant. He thought it could easily be a virus transplant, that the baboon would be donating not only its liver to the patient but a host of microscopic organisms as well. He didn't consider that science fiction; he considered it almost too real. Starzl might know livers, but Allan knew viruses. He'd been fighting against the AIDS virus for a decade.

By now, virologists were convinced that AIDS had begun in monkeys. In fact, when you really started looking at the plague of wild viruses carried by monkeys, it seemed almost miraculous that they hadn't been part of an epidemic before. Of all the animals that scientists use in research, monkeys pose the greatest risk to people. It's the same connection that makes them good transplant material, that they are genetically close. Viruses leap more easily from monkeys into humans because of that kinship. And monkeys are not "clean" animals. They come from the wild, bearing strange, microscopic creatures.

The very history of primate research, when you start looking for viruses, seems a history of near-misses and risk to the human population. At what point, Allan wondered, would we start learning from the past? At what point would transplant specialists question the wisdom of taking a wild animal's tissue, infested with mutation-prone viruses, and installing it into a human body? He had no doubt that baboon liver transplants—once successful, once survivable—could be the seed of that creeping epidemic that he paints so vividly. The infected patients, then their family, their friends, strangers on the street, an ever-growing chain of a sickness unseen before.

"I set that scene about 20 years from now," Allan says. "When baboon liver transplants are common. And these unexplained illnesses start occurring."

Maybe the possibility wouldn't have pulled at him if he wasn't an evolutionary virologist, toiling in the middle of the worst viral epidemic of our time, the relentless sweep of immunodeficiency viruses from monkeys to humans. Allan, raised in suburban Boston, became interested in animal viruses early. He has a master's degree in microbiology, a degree in veterinary medicine from Michigan State University. He had come to appreciate retroviruses in particular, as they ripped through animal species, causing leukemia in cats, cancers in cows and sheep. That such a tiny organism could manipulate the genes of such large organisms fascinated Allan.

He wondered how they did it. He wondered if, by flicking on the genetic switches, they could flick the wrong ones, triggering tumor growth. Researchers suspected that some viruses played a role in cancer. Perhaps these were among them. When he finished his degree in veterinary medicine in 1984, he took a postdoctoral job in the laboratory of Max Essex, a Harvard virologist. Essex was working with the most infamous of all retroviruses, the AIDS pathogen. It was incredible timing. Just as Allan arrived, the nearby New England Regional Primate Research Center (where Essex also worked) announced the isolation of Simian AIDS in its macaques. The next step seemed like simple logic. When his postdoctoral stint was up, Allan went to a primate center, Southwest in San Antonio.

"I'd already been collaborating with people here, doing work with HIV," he recalls. "But I felt that SIV was really the future, that we would understand the evolution and pathogenesis of the virus in that model." To be Jonathan Allan—to painstakingly tease apart the genetics of the AIDS viruses, to realize how dangerously intertwined the human and monkey viruses were, and *then* to learn that someone was putting a whole monkey liver directly into a human being—

was to be shocked. He could hardly believe it. His dismay went to war with his scientific ethic. He was a researcher. He wasn't a policymaker. He was a virologist, not a transplant surgeon. It was none of his business. And yet, and yet the potential epidemic kept creeping into his head: "If it wasn't for AIDS, maybe this wouldn't seem so dangerous."

"The people who are learning about AIDS are not the people making these decisions," Allan says in frustration. "The transplant surgeons are making the decisions, and they're learning, but not about AIDS. The first time someone asked me about this, I said look, it's not my area. But I've come to think that it's irresponsible not to think this through. To not talk about it when people are dying, it's not right. There's such a sensitivity among researchers not to talk about ethics, that if you do, you might be labeled one of the animal rights types. But I think it should be mandatory: If you do animal research, that you do it in a conscious way. If not, then people can say, look, it's just another jerk scientist [spouting] off the top of his head and not thinking about the consequences."

He wasn't alone in worrying about the consequences of baboon liver transplants. Out in California, at the Simian Retrovirus Laboratory, Nick Lerche was struggling with the same sense of dismay and disbelief. Like Allan, he'd spent the best part of a decade trying to figure out a way to push back the AIDS tidal wave. He picked up the local newspaper, read about the liver transplant, and thought to himself, "Wait a minute, this is right out of science fiction. If I was Michael Crichton, I'd be on my way to another best-seller."

It wasn't just the possibility of moving viruses from monkeys to humans that troubled Lerche. It was that the operation seemed designed to encourage it. The human receiving the liver would be loaded with anti-rejection drugs, designed to suppress immune function. The drugs were guaranteed to stop the patient's body from fighting off a strange new virus. "You could not design a better experiment of cross-species transmission of a virus than to transplant an organ from a baboon to a human, and then immunosuppress the hell out of the human to keep it from rejecting the organ," Lerche says. "Who knows what might happen?"

The path of infection runs two ways between monkeys and humans. A single person infected with tuberculosis, not even feeling sick, can bring down a primate colony. Tuberculosis, carried by an airborne bacterium, races through rhesus macaques and levels them. Primate centers have become meticulous about keeping TB out. Employees require regular TB tests. Visitors must be swaddled in masks, gowns, protective gear. Human childhood diseases, too, can be lethal in a primate colony. At the California primate research center, a lab technician came in one day accompanied by several children, one of whom was developing measles. The measles epidemic in the Davis colony killed almost 80 rhesus macaques. Measles in high gear wipes out the immune system almost as efficiently as AIDS, the difference being the system bounces back. The monkeys died of roaring opportunistic infections, especially pneumonias. "We were out there setting up field ICUs, reinflating monkeys, lungs," assistant director Jeff Roberts remembers. "We still couldn't keep them alive. It was definitely the low point in my career." He now has every monkey in the colony vaccinated for measles and maintains a unrelenting grudge against the 1980s Reagan administration for cutting back on federally sponsored vaccination programs for children. He sees those cutbacks as allowing measles to flare up again in the population, putting his colony at risk.

If you think of disease transmission as following an arrow, the measles arrow points human-to-monkey. There are other viruses, though, that point monkey-to-human. Like measles, when they hit their target, they are deadly. The classic one, feared among primate researchers, little known outside that circle, is an obscure infection called "monkey B" or "B virus." B virus is a member of the herpes family. It is almost a ringer for herpes simplex in humans, the infection that causes cold sores. In rhesus macaques, B virus produces an insignificant illness, usually causing nothing beyond the occasional blistered tongue or reddened eyes. Until the 1930s, no one even knew it existed.

They found out in the nastiest way, watching a colleague die. In 1932, a young physician—known forever in medical records as Dr. W.B.—was bitten by a macaque. Two days later, he was admitted to a hospital with numbness in his arms and legs. Two days after that, he was dead, his lungs suddenly stiff, foamy fluids spilling out of his mouth and nose. It didn't take a genius to connect the death to the monkey bite, but no one knew why. Enter Dr. Albert Sabin, a young scientist with a fascination with viruses. Sabin, 20-odd years later, would dazzle the world by developing the live-virus vaccine for polio.

Back then, Sabin and colleague Arthur Wright were fascinated by the monkey-bite death. It took them two years, but eventually they were able to prove that there was some kind of virus involved, something so tiny that it could slip through the filters designed to trap bacteria. It fit the pattern of a herpes virus, and Sabin was convinced it was a new one. He and Wright named it B virus, after Dr. W.B.

Herpes viruses are hard to isolate. They tuck themselves into nerve cells so discreetly that prying them out again is, still, a major task. It wasn't until nearly the 1950s that B virus itself was isolated from monkeys. By that time, polio vaccines were being dished out, grown in monkey cells. And suddenly, everyone realized that this potentially very deadly virus could be simmering along right in the vaccines. The thought that it was being accidentally injected into thousands of Americans was the stuff of nightmares.

It turned out to be a near-miss. Polio vaccines were routinely treated with formalin, a toxic solution of formaldehyde in water. Formalin was used for a reason. It was supposed to "purify" the cell culture, meaning kill every unwanted organism. Luckily it worked with B virus; it purified the B viruses in the culture right out of existence.

B virus still haunts primate laboratories, though. It remains a threat to anyone in contact with live monkeys or monkey-cell cultures. Since the death of Dr. W.B., 32 American lab workers have been infected by B virus, and 23 of them have died from it. People in labs wear masks, gloves, and booties, protecting their skin. B virus moves most easily through breaks in the skin—a bite, a puncture, a scratch. Despite those precautions, despite a general attitude of paranoia, the virus has killed 5 primate workers since 1987—2 in Florida, 2 in Ohio, 1 in Texas. They were bad deaths, the virus burning like fire up the spinal cord and into the base of the brain. Fire and heat are a good analogy for the destruction of B virus; what's left behind is blackened with blood. The autopsy slides of B-virus-damaged brains resemble the blasted-out look of a torched building—black, stripped to frame. In one of the Florida deaths, that of a 32-year-old man, pathologists found that his upper spinal cord had turned liquid, melted by that strange kind of viral heat.

Like the immunodeficiency viruses, B virus begins mild in the host monkeys and turns wicked in humans. Just as people were starting to realize that monkey-to-human transmission usually meant bad news for humans, they found yet another virus in the polio vaccines. This one survived formalin.

The troublemaker was a papovavirus discovered in 1961. Papovaviruses are offbeat, obscure viruses that no one understands too well. An obscure virus is a virus that isn't usually a problem for people. The ones that are—influenza, herpes, measles, HIV-1—are household terms. Who's ever heard of BK or JC virus? Those are papoviruses that occasionally show up in human tumor cells. Who had ever heard of SV40, the scientific shorthand for Simian Virus 40? That was the agent that popped out of polio vaccines, and it looked like trouble from the start. SV40 was a tiny virus that liked to tuck itself into kidney tissue, exactly the tissues where vaccines are made. It was a shifty, untrustworthy RNA virus. It slid into the center of a host cell and multiplied. It was that rapid integration into the host that saved it. Socked away in the cell, it was buffered against the formalin.

SV40 was the first papovavirus ever found in a monkey. It was a lousy way to find it, though. It appeared that while physicians were vaccinating the American population against polio, they were simultaneously—between the years 1953 and 1961—inoculating them with SV40. Even worse, the virus was infectious. A tentative, hasty sample of blood tests turned up many people who were antibody-positive for SV40. No one knows how many because the government didn't want to conduct tests. It wasn't B virus, after all. And nobody seemed to be dying from it.

Vaccine makers began destroying the virus in vaccine cultures in 1961, cranking up the chemical treatments. Still, they didn't toss the infected vaccine preparations. Apparently, physicians kept using them until 1963, when the SV40-contaminated stocks were used up. If the baby boomers had started dropping like flies, then the studies would undoubtedly have come. This virus doesn't seem to have killed anyone, or at least not many. No strange epidemic started rippling across the country. The missile that got through the researchers' defenses has been a dud, as far as scientists can tell. Doctors have isolated it—extremely rarely—in human tumors growing in the passages in the nose. They found it, once, in a case of rapidly spreading malignant melanoma. It has been extracted, also rarely, from the occasional tumor in rhesus macaques. There's no chain of cases, though, no hint of one strange cancer after another.

The remaining troubling question about SV40 is whether it poses the biggest threat to the children of the boomers, that in vaccinated-mother-to-child transmission is where the problems will come. A few studies have suggested that the children of polio-vaccinated mothers have a higher rate of cancer. That seems particularly true for nervous system cancers, brain tumors again. In the late 1980s, scientists tracking the life histories of 59,000 pregnant women, all vaccinated with the Salk polio vaccine, found that their offspring had a 13 times higher rate of brain tumors than those who did not receive the vaccine.

The very name SV40—the 40th monkey virus, in other words— makes it clear that scientists knew, back in 1961, that they had an onslaught of primate infections to worry about. There are now entire books devoted to the issue. Most of them run between 400 and 500 pages. The lists include more than 50 types of bacteria; more than 30 fungi; blood parasites; and microscopic worms such as the thread-like *Filaria*. Those tiny worms invade body cavities and reproduce until they can knot the tissues into misshapen lumps. A common name for one such infection is elephantiasis.

The parasites include malaria, toxoplasmosis (an infection, also found in cats and known to cause birth defects), and a disease of growing reputation, *Pneumocystis carinii*. Monkeys frequently carry the last, an organism that causes a devastating pneumonia, filling the lining of the lungs with thick, scabby cysts, rendering them useless. Pneumocystis pneumonia is one of the major causes of death in AIDS patients. Monkey-borne bacteria include tuberculosis, leprosy, common pneumonia-causing bacteria, tetanus, salmonella, listeria.

The viruses themselves are both countless and uncountable. Spread through the wild primate family are more than 35 herpes viruses alone. Viruses are the hardest to find and, because of that, the easiest to worry about. B virus was a surprise; SV40 was a surprise; SIV was a surprise. Virologists suspect that the animals carry still more unknown viruses, unknown only because people haven't figured out how to see them yet. Viruses figure into Allan's transplantation fears because as a class, viruses scare the hell out of any researcher who has studied them. They are too poorly understood, too elusive, too evil to treat casually.

Without even causing a single human illness, viruses in monkeys are capable of sending health authorities into a complete panic. It's a curious paradox: We import the monkeys; we use them in research; we transplant pieces of them into human beings. The justification is that we use them to halt and prevent human illness. All the time, when we bring the monkeys out of the rainforest, we are bringing their diseases with them and exposing people here.

The classic example of a near-miss along that line is the story of the Ebola-like virus that came into the United States in 1989, piggybacking in the blood of crab-eating macaques from the Philippines. To understand just how scared health officials were, exactly why the U.S. Centers for Disease Control virtually shut down the monkey import business, you have to appreciate what a vicious infection an Ebola virus can be.

Ebolas are filoviruses, named for their fine, thready appearance under a microscope. They look like gleaming strands of crystal, effortlessly elegant, endlessly deadly. You can say, calmly, that filoviruses cause hemorrhagic fevers. Or you can tell what happened in the Sudan and Zaire, in the mid-1970s, when suddenly, out of nowhere, Ebola viruses flared in the forest, leaving almost no one alive to tell the story. The death rate in some villages of the Sudan topped 90 percent. Ebola killed patients, their nurses, their doctors, wiped out entire hospital staffs. By the end, anybody left alive was fleeing the villages. The best thing you can say about an Ebola death is that it is quick. The worst is that it seems endless. The virus destroys the body's blood system, ripping apart blood cells until they leak as if punctured by a thousand tiny needles. In the end, the membranes that contain the body's fluid are destroyed. The body is awash in disintegrated blood, the skin mushy and oozing. An Ebola victim dies when he suffocates in that terrible internal flooding.

Only once had a filovirus been brought out of Africa in monkeys. That was a cousin of Ebola, called Marburg. In fact, the United States had no quarantine requirements for imported monkeys until after Marburg, sometimes called green monkey disease, emerged in Europe.

The common name comes from Marburg, Germany, where the virus suddenly flared in 1967. A shipment of African green monkeys had been imported there from Uganda. At Marburg, there was a processing plant, using monkeys in vaccine production. African green monkeys have long been used in polio vaccine work, particularly in Europe. At the plant, 31 people became sick; 8 died, a death rate of nearly 25 percent. That's far less than Ebola Sudan, but far higher than an infection like yellow fever, which kills about 5 percent of those infected. An investigation found that almost all those who died had direct contact with monkey blood. The victims were those who had conducted autopsies or surgeries. The only exception was a woman, the wife of one of the lab workers who died of Marburg. The virus was isolated in her husband's semen. In other words, the potential was there to slip the virus into the general population.

In the fall of 1989, crab-eating macaques suddenly started dying at Hazelton Research, an animal import facility in Reston, Virginia.

No one suspected anything like Ebola. The disease was so deadly, though, and so mysterious that Hazleton eventually sought help at a nearby laboratory, the U.S. Army's infectious disease center at Fort Dietrich, Maryland. Fort Dietrich is famous, or notorious, for biological warfare research. It had the best group of virus experts in the country, but even they didn't suspect Ebola.

Richard Preston, writing in *The New Yorker* magazine, recounts a story in which Dietrich scientists put the mysterious monkey agent into cell cultures, triggering a massive cell death. They speculated that they had contamination by a known cell-killing bacterium, pseudomonas, which releases a smell much like Welch's grape juice. The researchers then stood around, sniffing the rotting cells in the test tubes, trying in vain for the scent of crushed grapes. Two days later, when virologists spread the tattered cell culture under a microscope, they saw the crystal lines of an Ebola virus running through it, shining like ice beneath the lens.

Within days, CDC was running a monkey blockade around the country. Every monkey at the Reston facility was killed; in the cold jargon of medical bureaucrats, the center was "depopulated" of close to 400 animals. The agency sent out terse letters to administrators of primate facilities, beginning: "The health of your workers may be at risk because of Ebola virus infection." And it stopped the importation of macaques into the country. None of them would come through, CDC announced, until federal regulators had figured out how Ebola had slipped in.

As it turned out, Ebola Reston, as the virus was named, was something of a dud, too—a little like SV40. Except in this case it killed monkeys. CDC sent scientists to the Philippines, where they found a death rate topping 80 percent in some holding facilities for crab-eating macaques. Unlike its African cousins, though, this virus seemed reluctant to shift into people. Only two animal handlers, one in Reston, one in the Philippines, tested clearly positive for Ebola Reston. Neither became sick. The clearest finding by federal authorities was that the research community had gotten pretty relaxed about monkey viruses. CDC rushed teams of inspectors to quarantine facilities; and temporarily closed most of them down as health risks. In a single day in March 1990, the agency inspected eight facilities and revoked the import permit for seven of them.

Ebola Reston also proved once again that viruses baffle the best medical detectives. CDC researchers are still trying to figure out where that Ebola came from. It's not a long-existing monkey virus, such as SIV, or the monkeys wouldn't die so readily. Marburg was a monkey killer, too. Epidemiologists have looked at rats, bats, bedbugs, everything they can think of that breathes, and still don't know the carrier of the virus. They don't know where it will come from, when it will rise again, whether its next reincarnation will be something as horrible as Ebola Sudan. In a way, though, Ebolas are easy. They're like B virus, so deadly you can't miss them. Another retrovirus, another AIDS-like infection with a 10-year latency, something like that could be sliding on through, right now.

There are those who think that it was people's sublime ignorance about primate viruses—combined with a willingness to use monkeys as medical tools without question—that actually began the AIDS epidemic. One theory is that AIDS really began because of scientists blundering around, using monkeys in research without understanding the risks. Back in the 1920s, of course, before Dr. W.B. died, no one had made that connection. In those days, researchers freely transferred the blood and tissues of African monkeys directly into people. A classic example was malaria research of the time. In 1922, malaria researchers injected blood from chimpanzees and sooty mangabeys into human volunteers, trying to find out if primate malaria parasites could infect people. Dozens of people had primate blood dripped directly into their veins. Dozens more received shots of rhesus macaque blood, after the macaques were injected with mangabey blood. The experiments proved that monkey malaria did not endanger people; they had to be bitten directly by parasite-carrying mosquitos. As to whether the humans tested were infected with monkey virus, no one knows. Doctors at the time didn't even consider the issue or conduct followup tests.

In another set of experiments—clearly a product of the time—scientists grafted monkey testes onto older men. There was a theory in the 1920s that aging was due to a loss of testicular secretions. If you could restart them with young testes, you could stay forever young. That project—which had hundreds of volunteers—stopped in the mid-1930s when testosterone was isolated and shown to have no effect on aging. Either experiment might have been a jump-start for moving AIDS across the species line. Still, the human subjects were mostly American or European. Nothing in the spreading pattern of the diseases suggests that it evolved on those continents.

If AIDS did, indeed, spring out of African monkeys, then the one medical experiment which fits the pattern most closely is the polio vaccine program. In 1957 and 1958, the oral vaccine, grown in monkey kidney cells, was tested on several hundred thousand Africans in Zaire, Rwanda, and Burundi. Among the monkeys used to prepare the vaccines were African green monkeys, now recognized as a longtime carrier of SIV. The vaccine was spray-injected into the mouths of the recipients. If they had a cut, a sore, an abscess, the virus might have slipped into the humans' blood. Maybe. But the disease does not seem to have flared up noticeably in places where the vaccine was given. And although African green monkeys were used in general vaccine preparation, there is no clear evidence they were used in these particular spray vaccines.

If anything, the questions raised about the AIDS-polio link seem to suggest that this was another lucky miss, that once again researchers used monkeys in a risky way without doing any real harm. If everyone wasn't so frustrated over AIDS, perhaps the vaccine idea wouldn't have gotten so much attention. It did, at least, catch public interest. One of the publications to explore the idea was *Rolling Stone* magazine, in the spring of 1992. The research establishment journal *Science* responded with the sarcastic headline: "Debate on AIDS Origin: Rolling Stone Weighs In."

The sniping over whether polio vaccines had kicked off the AIDS epidemic was still going on when Starzl performed his baboon liver transplant in Pittsburgh. To Nick Lerche, in California, it was a shocking contrast—the near panicky discussion about whether a few monkey cells could have triggered AIDS in humans, the surgical implantation of an entire monkey liver into a human being. "It was so bizarre," Lerche says. "There was everyone up in arms about the polio vaccine and AIDS. And there were people putting a whole monkey liver in a patient. And yet it was like they weren't making the connection—that it was basically the same risk."

Lerche, of course, also lives in the world of primate research into AIDS. Transplant surgery is another world and the worst thing about it is watching people die. The numbers are dismal more than that, painful: about 4,000 human organ donors a year; 29,000 people on the waiting list for transplants, hoping that their hearts, lungs, kidneys, livers, whatever vital part is failing, hoping it lasts just a little longer.

Even before he came to Pittsburgh, Starzl clearly was a man unable to just watch people die. A brilliant, seemingly tireless surgeon, he pioneered human-to-human liver transplants, and by the end of the 1960s, he was making them work. He pushed the anti-rejection drugs. He was among the first to try an experimental drug from

Japan, FK506, so far, the best anti-rejection drug known. Without FK506, Starzl might not have tried the baboon liver transplants in 1992. The track record of cross-species transplants is dismal. In the last 30 years, Starzl and other surgeons have tried 30 times to make other animal's organs work in human bodies. Everyone has failed.

In 1963 and 1964, surgeons in Louisiana transplanted twelve chimpanzee kidneys into humans. The longest survival was nine months. In the mid-1960s, Starzl himself transplanted six baboon kidneys into humans. His patients lived between three weeks and three months. In 1964, a chimpanzee heart transplant—two-hour survival. A short time later, Starzl tried again, putting chimpanzee livers into children. He operated on three children; none survived longer than two weeks. In 1977, Christian Barnaard's two, much-publicized chimpanzee heart transplants failed—both within days.

In 1984, in California, Loma Linda University surgeon Leonard Bailey took the heart of a young baboon—also from Southwest—and put it into the chest of a newborn child. They called the little girl Baby Fae. It was something about that combination—the baby and the baboon heart. People watched and hoped. Baby Fae lived 20 days, killed by too many shocks—the strange heart, from a different body, a different species, even a different blood type. Her blood system rebelled at being pumped through that alien heart; the individual cells began thickening, clotting up, refusing to move. Bailey, to this day, keeps her old, bad heart floating in a jar of preservative on his desk, a reminder. He continues—as Starzl has done—experimenting with animals. He moves hearts between baboons and rhesus macaques, trying to make the one beat reliably in the other.

It was eight years after Baby Fae before anyone tried a baboon-to-human transplant again. On June 28, 1992, Starzl took the liver from a 15-year-old male baboon and used it to replace the dying liver of a 35-year-old man. The man had AIDS but he also had chronic hepatitis; people with that disease are crossed right off the human donor list. The hepatitis virus is extremely durable. It hangs on through the transplant; it can destroy the donated organ as well. The choice was the baboon liver or nothing. Starzl and his patient took the monkey transplant.

The Pittsburgh team's main fear was rejection and, as they admitted later, they loaded their patient with FK506, wiping out his immune response, preventing it from booting the liver. As a result, though, opportunistic infections swamped the patient; he had nothing to fight them with. When he died, on September 6, it was from a spreading fungal infection that ate through the walls of blood vessels in his brain. The baboon liver hadn't performed perfectly, there was a mucky sludge in its ducts, but it hadn't quit either. Pittsburgh had approved three more baboon liver transplants, and Starzl performed the second one a few months later, on January 10, 1993. That patient, a 62-year-old man, also was killed by a fast-crawling fungal infection, some four weeks later.

The operations are after all, as Baby Fae's was in 1984, an experiment—a deliberate crossing of the species lines. Nature holds those lines apart; you cannot mate a human and a monkey, breed a baboon with a pig. Yet baboon hearts and pig valves have been sewn into the human body. Transplant surgery is a breaker of the lines, using surgical techniques and modern drugs, to take down nature's barricades. Modern medicine is learning how the liver of a baboon can be made to filter the blood of a human being. Starzl has suggested that baboons are only the beginning. As doctors understand better how to handle rejection, find the right balance of drugs, they may try other species. "Cracking the species barrier," he calls it. "Whatever it takes to stop people from dying," said Starzl, at a University of Pittsburgh press conference. "They get our passion, they get our commitment. Our ultimate fidelity really has to be to our patients."

"Fidelity," "passion," "commitment"—the vocabulary of a rally for support, a drum beat of inspiration. Down in Texas, though, Jonathan Allan has successfully braced himself against the emotional power of that call. This isn't about emotion to him. When Starzl uses sweeping phrases like "cracking the species barrier," Allan tends to think, instead, about consequences. Cracking the species barrier means opening what shadowy Pandora's box of virology? "Starzl may say that he sees no real ethical issues here," Allan says. "But if you don't see them, then you're the wrong person to be making the decision. Maybe this is not an area to be decided by transplant surgeons at all."

There is some debate about whether the primate colony managers at Southwest were clearly told by Pittsburgh that it wanted to purchase baboons for liver transplants. Certainly, they knew that it was Starzl's group that was interested, leading to an obvious conclusion. Still, some staffers have complained that they would have chosen a "cleaner" animal if they had known exactly what the plans were. Veterinarians at Southwest said that the animal used for the first transplant, in 1992, was infected with at least three viruses: cytomegalovirus, which can attack the eyes and the heart of immune-suppressed patients; Epstein-Barr, suspected in cancers of the blood; and SA8, the baboon version of B virus in macaques.

The decision process and the operation set off a genuine ripple of alarm in the virology community. There was an unusual flow of criticism. The fundamentally pro-biomedicine publication, the *Journal of NIH Research*, carried a review of the issue in which virologist after virologist worried about the risks of moving an infection across species lines. Among them were Jonathan Allan, Ron Desrosiers at the New England Regional Primate Research Center (one of the first to identify SIV in macaques), and Stephen Morse, an expert on emerging viral diseases at Rockefeller University in New York. Morse cited a litany of animal viruses transferred to humans—influenza, which begins in birds; Lassa fever, carried by rats; Korean hemorrhagic fever, carried by field mice; Marburg, Ebola, AIDS.

In the same article, John Fung, a member of Starzl's transplant team, pointed out that baboon livers may be the only answer for people suffering from hepatitis B. The monkey organs appear resistant to the human virus. Fung, as Starzl has done, emphasized the group's dedication to saving people. It's important to emphasize that no one has accused Starzl's group—or any other scientist interested in cross-species transplants—of careless work or immediately endangering public health. The discussion is about possible risks and how to respond to them. Scientists, such as Allan, do not demand that the operations be stopped. Rather, the virology community has both acknowledged the pressure to save human lives and argued for a voice in the debate over doing cross-species transplants, for a recognition of its concerns.

The University of Pittsburgh team has also made it clear that it considers primate viruses an issue to be handled carefully. After the two transplants, the university made a point of explaining that it did screen for viruses. Anyone receiving a baboon transplant would go immediately onto antiviral drugs. Any baboon considered transplant material would first be placed in quarantine, two independent laboratories would screen its blood for retroviruses, herpes viruses, hepatitis viruses, parasites such as *Toxoplasma gondii*, a blood parasite, in all more than twenty potential infectious agents. Any animal that was infected with a retrovirus would be taken out of the transplant program, with the exception of a class of retroviruses called foamy viruses, associated with no known disease.

The Pittsburgh scientists also decided to take a serious look at live viruses in captive baboons. They chose as a collaborator one of the country's top primate virologists, Seymour Kalter, a former researcher at Southwest who now runs his own business, the Virus Reference Laboratory. With Kalter's help, the Pittsburgh scientists did a viral hunt in 31 baboons. They were—a little dismayingly—successful. Thirty of the animals carried cytomegalovirus, SA8, and Epstein-Barr. Eight were infected with pathogenic retroviruses. Three were hepatitis carriers; ten had the toxoplasma agent in their bloodstreams. None of them carried Marburg, or monkey pox, or Encephalomyocarditis virus—the last an organism capable of chewing apart the heart muscle, and responsible for some massive baboon losses at Southwest. Thirty of them though—97 percent—were infected with foamy viruses.

Their approaches to baboon foamy viruses emphasize the difference between Allan and Starzl, with Starzl's focus on the suffering of the individual. Allan works in a primate center, where staffers are trained to guard themselves against infections carried by primates. It's not even the super-hot viruses, like Ebola. Places that handle the Ebola class of viruses become viral prison camps. You can pick out a hot-virus building by its paramilitary look. The special pathogens lab at CDC Atlanta is guarded by motion sensors, security guards, surveillance cameras. The cameras perch on high black poles, turning their glassy lens back and forth, back and forth ceaselessly. No one goes into a room with live Ebola without wearing a space suit, breathing through an air hose hooked to a nozzle in the ceiling, gloved, booted, wrapped. Even with all those barriers, no one leaves without standing in a five-minute Lysol shower. No one gets out the door without using a key card and an individual code. At Southwest, where the big risks are retroviruses and herpes viruses, researchers must wear gloves and masks. They are required to anesthetize an animal to be handled, for fear of being bitten.

From the Pittsburgh perspective, the approach to foamy viruses is a little more detached. They're retroviruses, the same class that's produced AIDS. But they seem to be sleepy little cousins of the killers, like Ebola-Reston rather than Sudan. You can find them but they seem harmless. A transplant surgeon, weighing the possible risk of a so-far benign virus in monkey tissue against the possibility of saving a dying patient, may well decide that he must try to preserve the human life. That pioneering medicine cannot be stopped by fears of the unknown. From the Southwest perspective, where staffers are taught to respect and fear primate viruses, it's impossible to dismiss any monkey retroviruses, even the seemingly benign foamy virus. In the back of his mind, Allan remains on foamy virus watch. He knows that scientists across the country are trying to figure out what foamy viruses do. They belong to a troubling family; because they are cousins to the AIDS virus, somehow viral specialists are unable to convince themselves that foamy viruses are harmless.

The kinship between foamy viruses and immunodeficiency viruses is most clear in the genetics. Both are RNA viruses, dependent on the genes of the host cell to function. Both weave themselves into the host cells. Both stay there a long time, perhaps a lifetime. It would be easier for Allan to accept a gamble on foamy viruses if AIDS hadn't flashed out of nowhere barely a decade ago. He can accept that foamy viruses are mysterious, but not that they are meaningless. People might have said that about the AIDS virus, back in the 1970s, before the inexplicable illnesses began appearing. AIDS, after all, can wait quite a while before flying its flags, showing symptoms. The latency period—infected but asymptomatic—can last years, by some estimates more than a decade.

If virologists take baboon tissue and try to isolate SIV out of a blood-cell culture, the foamy viruses don't look harmless at all. Sometimes it's hard to get the AIDS-like virus out because the foamy viruses kill the cells first. "The foamy virus just comes up so fast," Allan says. They're called foamy viruses because of what they do to the cells, exploding them into bubbling, balloon-like structures called giant cells. Giant cells look like foam; they are the signature of a cytopathic virus, a cell-killer. B virus from macaques can turn human cells into a bloody explosion of giant cells. The bubbling of cells caused by AIDS viruses duplicates foamy virus destruction.

Virologists just keep poking at foamy viruses, hunting for that undiscovered link to illness. At best, they've gotten vague hints. Traces of human foamy viruses have been found in the white blood cells of people with Grave's disease, which damages the thyroid. If transgenic mice—mice implanted with human immune systems—are infected with human foamy viruses, the animals suffer nervous system damage. In 1992, researchers isolated a new foamy virus from an orangutan with a fatal disease, characterized by inability to control muscles, wasting of nerves. The ailing orangutan's problems looked a lot like the nerve problems found in the foamy-infected mice.

It's that shadowy pattern of nerve destruction that underlies Allan's vision of an epidemic brought on by monkey liver transplants.

He bases it on the nervous system damage already linked to foamy viruses. There's one other characteristic of foamy viruses that makes them a likely suspect in an epidemic. They appear—among monkeys—to be an airborne infection, carried by breath or touch, invading through the respiratory tract. AIDS, by all accounts, is a difficult virus to catch, transferred by infected blood or fluid. Simian foamy viruses—there are some 15 known including the baboon variety—appear to move around more easily. A retroviral disease blown about in the air could be a fearful prospect.

"People keep looking for the disease and we'll find it eventually," Allan says. "The transplant people say to us, you can't even tell us what the risk is. We can't be paralyzed by something you don't know. And the answer is, there's so many examples of viruses coming from one species and getting into another and causing a horrible disease. And you are coming up with the best scenario to do that, which is to take a whole organ out of one species, and put it into another. And even if you weren't immunosuppressing the hell out of it, you're still talking about a whole organ with all the viruses and bacteria and weird stuff, and putting it into a new environment. You've got a virus in a strange place. It's trying to figure out how to get into the strange cells. And maybe it goes into the human cells and maybe it comes out a different virus. That's the worst case, I think. That out of this we get a whole new virus."

The trouble with retroviruses is the way they make themselves part of the human cell, weaving into the DNA, turning genes on and off like a drunken piano player trying and failing to tap out a tune without errors. Wrapped into the genome like that, they can become integrated into the body—passed from parent to offspring. Researchers call that kind of sleepy genetic transfer an endogenous infection. Such viruses can be invisible, never causing an illness, wholly integrated into the cell. If you worry about them at all, you worry that they are still a kind of quirk in the genetics. Because of them, someday, the wrong gene may turn on—the same risk linked to vaccinating with a partially disabled, live AIDS virus.

Virologists know that a new environment often jump-starts viral change. Strange virus, strange host, unpredictable genetics—what will you end up with? It becomes a kind of viral roulette. Maybe nothing, and maybe a virus that a science fiction writer might dream up on a dark night. Seymour Kalter himself, down in San Antonio, did the classic experiment along those lines and the results were so unnerving that other scientists asked him, publicly, to stop.

Back in 1976, Kalter was wondering about primate retroviruses. One group of them—called type C retroviruses—seemed to be a weird mix of dangerous and dull. They were latent infections, living in monkeys, doing nothing. Then all of a sudden, their RNA could be isolated from monkey cancers—leukemias in gibbons, tumors in woolly monkeys. Kalter speculated that the C viruses might interact with something—another virus, a toxic compound—to become more dangerous. Kalter, called Sy by his friends, is a tough, independent-minded virologist. He left Southwest, in part, because he wanted to work for himself. He has a reputation for speaking his mind. He's been one of the most vocal critics of CDC's panicky handling of the Ebola crisis. He likes hard questions.

And the question that nagged at him about primate retroviruses was a what-if question. What if you allowed a merger of a primate retrovirus with a virus from another species—would you get a new virus? Would it be a dangerous one? Kalter infected a batch of human cells with a cancer-causing mouse virus, the murine sarcoma virus. He then mixed them and grew them with another line of cells, dog thymus cells which had been infected with the baboon type C retrovirus. The result was a new type of organism, a hybrid of the baboon and murine sarcoma viruses. It was one of those times when you want to put the genie back in the bottle, and quickly. They had created a cancer-causing virus that did, indeed, break the species barrier.

You could inject it, as Kalter did, into chimpanzees, beagles, marmosets, crab-eating macaques, and even baboons. Every one of them developed cancerous tumors. If so many species were vulnerable to the virus, then clearly people might be as well. The immediate fear was: What if this virus got loose? In a letter in *Science* magazine, other cancer researchers warned that the production of malignant tumors in so many primate species strongly suggested that humans could also be at risk. The letter said, in part, "We urge that all experiments involving co-cultivation of oncogenic [cancer-causing] viruses with primate viruses, be extensively evaluated."

Kalter did stop his baboon-mouse virus experiments. Still, to a virologist, a baboon liver transplant is, in some ways, just a large-scale version of his 1976 experiment. "Bingo," says Allan, when asked about the connection to Kalter's study. Kalter, too, suspects that mixing primate viruses and human genes is a gamble, weighing a single human life against the chances of widespread infection.

The simplest solution to fears about transferring infections from research primates would be to raise exceptionally clean colonies of animals. The National Institutes of Health has begun to fund Specific Pathogen Free (SPF) colonies of monkeys, carefully selected animals who seem to appear free of certain infections. Not that an SPF colony means a germ-free animal. It may mean B virus free, maybe, but that still leaves several hundred other likely organisms. "You can say, let's screen for Marburg, and declare your colony SPF, free of that virus," Allan says. "It doesn't necessarily mean anything. It just makes people feel better. It's like saying we're going to screen the viruses out of the baboon before we do a transplant. It's absurd to even talk about it."

Suppose, he suggests, that the baboon transplants had become successful in the 1970s? Who would have screened for immunodeficiency viruses? No one knew they existed. The chances remain high that there are other primate viruses, as yet unknown. And there's another chance to consider. That tomorrow, someone discovers that foamy viruses really do cause something like that strange, wasting disease that Allan thought up in a dark corner of his mind.

SPF

Specific Pathogen Free

ELEVEN

The Last Mangabeys

THE TULANE REGIONAL Primate Research Center specializes in infectious disease: AIDS, filariasis, leprosy, malaria, tuberculosis, Lyme disease, hepatitis, Epstein-Barr. The list scrolls on like a roll call of criminals, a rogues' gallery of Most Wanted Organisms reminiscent of the smudgy photos of murderers tacked into the corner of a post office.

In any shape, a killer is a killer, chilling and oddly fascinating. Standing in front of a murderer's photo, wondering what made him so deadly, is not so different from staring at the outline of a virus or a parasite, wondering what in that incredibly tiny package makes it lethal. The survival trick is, with either, to keep your distance. Tulane malaria expert Frank Cogswell once became so entranced by the parasite that causes malaria, he almost let it kill him.

Cogswell has carried on a highly personal war against malaria since the early 1970s, when he was a Marine platoon commander in Vietnam, helplessly losing more men to the disease's relentless fevers than to battle-fire. When the war was over, he returned to school and majored in tropical medicine at Tulane. He later spent three years in Cameroon studying malaria. In the third year, 1987, he caught the disease. Recognizing the early symptoms, he drew some of his blood, smeared it across a glass slide, and slid it under a microscope. There, magnified more than 1,000 times, was the round outline of the plasmodium parasite, bearer of malaria. It was pretty sparse though. The infection would be more interesting, Cogswell thought, if he let it go a little longer.

Now, he can smile, shaking his head, admitting, OK, he'd gotten a little cocky. There's a relief in sitting here, the Louisiana sunlight glinting off the glass of his office windows, the rough carved masks from Cameroon (the ones his wife refuses to let in the house) frowning down from the walls around him. Back then, for a while, he just figured he'd be buried in Africa. "I looked at the blood and it was less than 1 percent infected," Cogswell says. "I thought, it would be a much better teaching tool if I let it go a cycle or so longer. So I let it go, and I got so sick that I passed out. I had a 105-degree fever, sweats, chills. I thought it would be easy to treat myself. I couldn't even keep the chloroquine down. And talk about hallucinations. I was ready to die when the fever broke."

He hasn't lost his interest in malaria. Due to both insecticide resistance by mosquitos and new drug-tolerant strains of the malaria parasite, the toll has been rising worldwide. Some 200 million cases are now reported yearly, mostly in tropical countries, and more than 2 million people, mainly children, die from malaria annually. Still Cogswell keeps his careful distance now, studying the disease in rhesus macaques. He works with four young monkeys at a time, injecting the malaria parasite through their ears. It may take repeated injections for the parasites to flourish in the animals' blood, to multiply as they did in Cogswell's body. When the monkeys develop an antibody response to malaria, Cogswell returns them to the general breeding population. "I give them malaria and I cure them," Cogswell says. "I don't let them get very sick before I bring out the drugs, because I remember what it's like. And I don't quit because I also remember that children are still dying of this."

Usually, researchers remember not to dance too closely with a killer. They put on the space suits at CDC; they draw on the gloves at Southwest; they remember that the classic human model of a tropical infectious disease is often a corpse. They stand back, let the infection simmer in monkeys. Down the tiled hall, around a corner from Cogswell, Tulane biologist Robert Gormus has developed his own fascination with leprosy. Leprosy is not a disease for scientific gambling. It's destruction is too visible, too rotten. And although it is a minor threat in this country, some 12 million people around the world have been infected with the disease.

Leprosy comes by a mycobacterium, related to the one that causes tuberculosis. The trick about leprosy is that the bacterium likes the cooler tissues of the body—the skin, the surface nerves, the eyes. It spreads there like mold on bread, a kind of creeping invasion. The leprosy bacterium destroys what it invades. It deadens the nerves, numbs the muscles, breaks down the skin. In the Middle Ages, people with leprosy would wear hoods, casting black, concealing shadows over what remained of their faces. "Monkeys are not as vain as humans though," Gormus says, half-jokingly. "I see male monkeys mate with females whose faces are eaten away by the disease. I don't think you'd see that with people."

For all the centuries that leprosy has plagued humans, no one has yet figured out how it works. How does that little bacterium knock down the immune system so easily, opening up the body for invasion? That was exactly the angle that interested Gormus so much. How did the leprosy bacterium squeeze past the body's defenses? If he could figure out the immune side-step, he thought, then he might block it. He might find a way to make a vaccine against leprosy, trick the body into fighting off the infection. If he had monkeys, the right monkeys, Gormus thought, he might be able to not just decipher the disease but prevent it.

Researchers had learned in the early 1980s that African sooty mangabeys, like people, sometimes became naturally infected with leprosy. The disease in mangabeys ran an eerie parallel to the human one, disfiguring but not deadly. So Gormus wanted mangabeys for his study. If he'd been able to simply use rhesus macaques, as did Frank Cogswell, it would have been just another project, another interesting foray into killers' row. With the mangabeys, though, everything changed.

In the end, that fact, just the basic need for monkeys, would cause Gormus almost as much trouble as malaria ever caused Frank Cogswell. A scientist specializing in infectious diseases expects his biggest risks to be the accidental needle stick. In the course of trying to get monkeys for his study, Gormus would spend nights expecting to be hacked to death by animal smugglers, sitting nearby with their machetes. He would rattle around the back roads of Africa in an ancient BMW with a cracked axle, driven by a bird smuggler with a penchant for running over every stray animal on the road. He would wonder if he would ever get home.

Gormus speaks in the warm drawl of a born Southerner and a born storyteller. The words flow, sometimes profane, sometimes soft as a Louisiana summer day, painting pictures of Africa, distant in miles and immediate in his mind. Those rocketing drives and the terrified animals, scrabbling in the road to get away. "I started grabbing the wheel away from him whenever we saw a dog, and he'd be screaming at me swearing, and I'd be screaming back, 'Shut the fuck up!' " Gormus says. He shakes his head: "It's an incredible place, a different place. Every time I went, I'd swear I'd never return. And then, I'd get back here, and starting dreaming about going back."

What happened to Robert Gormus, though, is more than just an adventure in Africa. It's a story of the growing dilemma for primate researchers, caught between research needs and the plight of primates in the wild. Gormus's story is not typical; most researchers do not end up driving around Africa fighting bird smugglers for the wheel of the car. That Gormus was forced into desperate measures, though, offers a rare look at what's involved in getting monkeys out of the wild for research. His plight stemmed from swelling environmental pressures and the disappearance of monkeys needed for research. Primate researchers are being pushed to find a balance between using primates and preserving them. Their conflict mirrors the global one—we are all caught these days between vanishing species and growing human needs.

If it was simply a matter of monkey bodies, it wouldn't be a problem. The basic monkey version of a laboratory rat, the rhesus macaque, breeds easily in captivity. Some 3,000 macaques are born yearly at the country's primate centers, such as Tulane. The centers have little need for imported macaques. According to NIH, which provides core funding for the seven regional primate centers, the operation is about 92 percent self-sufficient. The dilemma really comes with that other 8 percent. It represents a kind of open window, admitting a variety of species. There's no way to have a healthy breeding colony of every possible wild monkey. So if a scientist discovers that a rarely used monkey is a dream model for a certain disease, federal health authorities want the opening for imports.

The rare black-crested macaque of Sulawesi has been a prime model for diabetes; the threatened cotton-top tamarin, a small South American monkey, is considered an unusually good model for cancers. There's the pigtail macaque and its sudden promise as a model for HIV-1. In Gormus's case, there was the sooty mangabey and the stubborn and elusive leprosy bacterium.

The American monkey supply system, in general, buffers researchers from the realities of monkey-trapping in the wild. If you want one monkey or a dozen, you call the big dealers. In this country, they are Charles River Corp. owned by Bausch & Lomb; Hazleton Research Inc., headquartered in Denver; Buckshire Corp. in Pennsylvania; Worldwide Primates of Miami; Primate Products, based in the suburban sprawl south of San Francisco. The importers send you the animals, tested in quarantine, neatly boxed. You pay, of course. The bigger the animal, the higher the price. Crab-eating macaques, the choice of private pharmaceutical company researchers, run about $1,000 to $1,200; baboons, from $900 to $1,500; little South American squirrel monkeys and capuchins, about $800. The hottest monkey going—the Indonesian pigtailed macaque, with its potential as a model in the AIDS virus—runs about $1,500. Boxes cost extra.

It's as neat and trauma-free as ordering dishes from Crate and Barrel. It's what Gormus expected when he needed to acquire mangabeys.

In the beginning, the early 1980s, when Gormus first began working with leprosy, he was collaborating with the Atlanta-based Yerkes primate center, which had a longtime colony of about 150 mangabeys. The Yerkes colony was the only breeding group of mangabeys in a research center. Health officials were reluctant to have Gormus depend on—and diminish—their sole mangabey colony. Further, they thought it might be cheaper to have Gormus and his monkeys in one place, eliminating the shuttling between Louisiana and Georgia. In 1988, they agreed to fund the grant if Gormus could get his own monkeys.

That turned out to be not a little problem but a big one. Two years earlier, the U.S. Fish and Wildlife Service had classified the white-collared mangabey as an endangered species. Sooty mangabeys are so closely related to the white-collared mangabey that, for many scientists, they're the same monkey with slightly different fur markings. Gormus wrote to Fish and Wildlife to find out its position on the sooties. Today, he considers that the first step in an experiment gone haywire. "We foolishly, looking back, asked them their opinion. They first said they didn't know, and that we should back off until they made up their minds. Then they decided that the sooties were a subspecies of the white-collared mangabey," meaning they were endangered, too. And then, Fish and Wildlife said no; no endangered monkeys for leprosy studies. No monkeys balancing on the brink of survival could be imported for medical research. Hands off, please, Dr. Gormus.

Wild monkeys come from the tropics, mainly the rainforests and savannas of Latin America, Africa, and Asia. As the human population expands and human demands for resources grow with them, monkeys have found themselves sitting on valuable real estate—land people want. People need the acreage for farming. They clear it for the value of its trees or to make a space for their own homes. Four hundred years ago, tropical forests covered some 10 million square miles; now they cover fewer than 4 million. Conservationists say that about 50,000 square miles of tropical forest are destroyed each year, taking with them an astonishing array of animals and plants. This vanishing habitat, a mere 6 percent of the Earth's surface and shrinking, sustains an estimated 90 percent of the planet's species.

The crab-eating macaque is one of those species, by nature a forest dweller, a resident of the thick-packed trees along the edges of oceans and rivers in Indonesia. The last complete conservation surveys, done in the late 1980s, found that crab-eating macaque habitat had fallen by 66 percent in Sumatra; 96 percent in Bali. The white-handed gibbon, one of McGreal's pet species, once swung through 68,000 acres of lowland forest in northern Sumatra. That forest had been cut back to less than 30,000 acres, a 55 percent reduction. Many tropical countries have moved to protect their vanishing species, establishing wildlife preserves, protective agencies, and trying to halt trafficking in rare animals. Yet, they are often limited by simple finances. In the late 1980s, Sumatra had been able to set aside just 4,000 acres of gibbon habitat as a reserve.

The people-versus-animal debate plays out on a different level in rural Africa and Asia. It is not an ethical question as often as a question of survival. Who gets the food—hungry people or hungry monkeys? Intellectual debate on endangered primates is the luxury of a wealthy country. In the high mountains of the Philippines, there are villages that raise money by trapping monkeys for research. One American monkey importer tells of visiting one such tiny cluster of houses, where the custom was to name children only after their first birthday. So many babies died before the 12 months passed that it was better, less painful, to leave the infants anonymous.

Macaques can be fierce competition for food. The adventurous macaque family—especially the rhesus, crab-eater, pigtail, and stumptail—has a talent for thievery. They can strip a struggling family's garden bare, and they will. Baboons, too, are viewed as competitors, classified as vermin in central Africa. If they aren't competing for food, they *are* food. "Monkeys and manioc, monkeys and manioc," sighs Frank Cogswell about menu choices in Cameroon. In Liberia, Gormus discovered that people particularly liked to eat smoked mangabey. They would stack mangabey bodies like leaf piles on fires, first cutting off the inedible tails. "They didn't even gut them," he says. "You wouldn't believe how their bodies would puff up." Of the mangabeys he eventually did bring back from Africa, three he found in restaurant kitchens. He had to barter with the cooks to get them. In rural Asia, notably China, macaques are a popular choice for stew meat. People shoot the monkeys or poison them. One of the kinder ways they treat them may be to trap them and sell them for use in a research institution.

"It's all interwoven," says Shirley McGreal, of the International Primate Protection League. "You can't separate out the research demands from the logging, from the pest-shooting, the pet trade. It's a cumulative effect. What bothers me about the American researchers is that they know how bad the situation is. They know these animals are disappearing. It seems to me that we've reached a point that scientists should be asking the Kennedy question. Not what the monkeys can do for them. But what they can do for the monkeys, to stop them from disappearing."

The official total of wild monkeys imported is about 40,000 a year worldwide, with the United States taking most of them. In 1989, for instance, the United Kingdom imported 5,027 monkeys; Japan imported 4,702; the Netherlands 4,183; and France, 2,430. The United States brought in 26,479 monkeys. The numbers, everyone agrees, underestimate the actual take from the forest. By a lot. Trappers tend to seek out young animals; they are easier to handle and smaller, easier to ship. That means trappers often have to kill parent monkeys to get the young. They shoot the mothers out of the trees and grab the babies after they fall. They pry the youngsters away. The monkeys are carried away in cages, packed into shipping boxes, air-freighted to industrial countries seeking them for research use.

It's anyone's guess as to how many monkeys die in the long journey between forest and lab. Indonesian animal dealers have estimated that, of monkeys caught, between 32 and 71 percent die being carted out of the jungle or in holding facilities awaiting transport. Robert Gormus recounts being caught between two rival animal dealers in Sierra Leone. One man had thirty-eight mangabeys at his home but he wasn't Gormus's primary supplier. The dealer he was working with carefully forgot to tell him about the waiting monkeys. Finally, the rival became frustrated and left on another trapping mission, leaving the monkeys abandoned in their cages. They died, all but eight, of starvation. Another contact of Gormus's heard about the animals and scooped up the survivors. He tossed the animals into the trunk of his car and drove to Gormus's hotel. The trunk lid was thrown triumphantly open. Inside huddled four live mangabeys and four corpses.

In the holding facilities themselves, the main difficulty is infectious disease. As CDC discovered during the Ebola outbreak, the death rate among crab-eating macaques in some Philippine holding stations topped 80 percent. Beyond that, ailing monkeys often die en route to their destinations in America or elsewhere. Shirley McGreal's group, in the early 1980s, tracked the shipment of about 2,000 African green monkeys from Ethiopia, Kenya, and Somalia into the United States. Fish and Wildlife reports indicated that, of those monkeys, 250 were dead on arrival; 327 died in quarantine. All told, more than a fourth of the monkeys died before reaching a laboratory. More recently, in the summer of 1992, Miami animal dealer Matthew Block's company, Worldwide Primates, shipped 110 crab-eating macaques from Indonesia to Miami. All of the animals were dead on arrival, apparently from a stress-induced pneumonia. Twenty of the dead were infant monkeys. The typical conservationist's estimate is 5 to 10 monkeys dead for every wild survivor that arrives at a research laboratory.

It used to be worse. It's been a long time, but even primate researchers still talk with awe, and some dismay, about how many animals were used to develop a polio vaccine. "We went through a hell of a lot of monkeys," says one high-ranking administrator at the NIH primate program. Before the race for the polio vaccine, there were an estimated 5 to 10 million rhesus macaques in India. During the height of the vaccine work, in the late 1950s and early 1960s, the United States alone was importing more than 200,000 monkeys a year, mostly from India. By the late 1970s, there were fewer than 200,000 rhesus macaques in India.

The Indian macaque population has risen steadily since then because India banned export of macaques. Several years later, Bangladesh also cut off the sale of its macaques to foreign laboratories. Both governments acted partly because of disappearing monkey populations and partly because U.S. laboratories were using the monkeys in nuclear weapons studies.

The export bans forced the federal primate centers to become breeding colonies. Until then, NIH had taken the position that it was in the business of research, not animal husbandry. But with supplies imperiled, the agency adapted, setting a target of 85 percent self-sufficiency by the mid-1980s. Then it surpassed that goal. The cutoff of macaques forced another change as well. The American biomedical community began to look at other monkeys besides rhesus macaques.

They stayed with what they knew. First choice was the most rhesus-like monkey available, crab-eating macaques. These days, almost three-fourths of the monkeys imported into this country are crab-eaters. Like rhesus macaques, crab-eaters are considered relatively abundant. The lesson of India, though, has stuck. The government of the Philippines, the country that is the primary source of crab-eating macaques, announced recently that it would cut off export of wild monkeys by the mid-1990s, before the population founders.

An ongoing debate over the use of pigtail macaques in AIDS research clearly shows the tension. By the time the medical community became interested in pigtails, conservationists were primed. They saw a clear parallel between the pressure to stop polio and the pressure to block the AIDS epidemic. Polio had decimated the rhesus macaque population of India. Activists were not willing to let that happen again in another species.

Further, pigtails were never as abundant as rhesus. They were not starting from a population base in the millions. Estimates ranged between 500,000 and 900,000 pigtails in Indonesia, mostly living outside protected reserves. So people like Shirley McGreal were prepared to be very edgy about increased use of the monkeys.

In 1992, Indonesia exported about 1,200 pigtail macaques to the United States. Nothing compared to the polio drain on rhesus. Still, conservation groups were quick to hit the numbers. The export might appear small, but during the 1980s, only about 200 pigtails a year were recorded coming into the country. McGreal argued that the taking of the wild monkeys was pointless, pure insistence by scientists on having whatever they wanted. Why start yanking monkeys out of the wild when it wasn't clear that the pigtail model for AIDS was even worthwhile? The Seattle primate center, after all, was already dominated by pigtail macaques. Its colony included only 84 rhesus, 319 crab-eaters, but nearly 1,300 pigtails. Indonesia, in fact, has indicated that it shares those concerns and plans to limit exports of wild-caught pigtails.

At NIH, though, the approach is pragmatic and focused on human needs. AIDS is a killer of people. To refuse to pick up a weapon against it would be like a refusal to fight. Seattle might have a solid colony of pigtails, enough for exploration. But, if the animals could really help bring AIDS down, a lot more would be needed. Perhaps no one expresses that better than Milton April, a Georgia-born biologist, who directs the chimpanzee breeding programs and helps coordinate animal models for AIDS. April works in the National Center for Research Resources, which oversees such resources as live monkeys. The federal primate centers are run through that division. It's dedicated to one theme, to supply scientists with the tools that they need. In this case, the tools happen to be monkeys. NIH provided the Seattle scientists with $200,000 to expand the supply of pigtail macaques.

"Think about what we do here and underline the word resource," April says. "We aren't just sitting around scratching our heads and dreaming up uses for animals. We listen to our investigators. In the case of the pigtail macaque, we've got a major disease on our hands and an animal that, given time, could be the premier model for that disease. There's no way that we wouldn't consider providing those animals. If we err, we'll err on the side of helping build up another captive breeding colony of pigtail macaques. A lot of money is being poured into AIDS research; we can't afford to sit here with a responsibility to try to support the research and not do it. We'll take the criticism and the flack, and we'll make mistakes, but, by damn, we'll provide the resource."

And that was also the attitude of Robert Gormus when he began his mangabey hunt. He went in stubborn, determined to get the monkeys. He was able to rally enough political pressure—from legislators and NIH—to get his permit, a special exception to the restrictions on importing mangabeys. It was one of those decisions that satisfies no one completely. It was a yes, you can have the monkeys—so McGreal didn't like it. She thought researchers should stay away from endangered species. Gormus didn't like it either. There were strings, expensive ones. By condition of the permit, NIH had to spend $250,000 to survey the West Africa mangabey population; Gormus had to set up a mangabey breeding colony in Louisiana.

Still, he had the paper, the official okay to bring in 150 mangabeys between the years of 1988 and 1991. Finally, he was able to make that simple phone call to a dealer. He called Worldwide Primates in Miami, where owner Matthew Block had a long and friendly association with Tulane. Block was also the big dealer who did the most work in Africa. The problem was, Worldwide didn't have any mangabeys. Neither did any other dealer. There was Gormus with a million dollar grant and no monkeys. The choice was either study another disease or get the monkeys himself.

He wasn't naive enough, though, to think it was a matter of boarding an international flight, whipping out his checkbook, and buying monkeys. Everyone at the Tulane center, Gormus included, recognized that it was a risky venture. None of them knew how to find monkeys in the wild. If Gormus was going to Africa, he needed expert help. The center decided to ask Matt Block to at least point them in the right direction. Block agreed to both introduce them to some suppliers in Africa and act as banker. He helped them sort through government regulations and export documents. Even so, getting mangabeys out was so difficult, Gormus says Block's cooperation was essential to the project. That was one reason why, after the animal dealer was charged with involvement in an orangutan smuggling scheme, primate center scientists remained among Block's strongest defenders.

The most damning aspect of the orangutan trafficking charges was the underlying message: that importers could use their legitimate research business as a cover to trade in endangered species. This was not a connection that scientists really wanted. Block first attempted to plead guilty to misdemeanor charges. McGreal organized a letter-writing campaign calling for the stiffest penalty possible. There were familiar voices in that campaign—Jane Goodall, Roger Fouts, Prince Philip of England, who has spent years working with the World Wildlife Fund. There were also surprisingly angry voices from within the science establishment. McGreal was especially startled and pleased by a letter from Robert Sapolsky at Stanford University. Sapolsky, who does primate research both in Africa, in the field, and in his laboratory in California, had often spoken against animal advocates. He said as much in his March 1993 letter: "I passionately defend the need for research with, and the killing of primates in the name of biomedical research, and I have often battled with animal rights groups over this issue." Still, he wrote, this time he supported groups like McGreal's in their insistence that Block not be allowed to plea bargain away a jail sentence.

"We scientists are often castigated by animal rights groups for the pain that we inflict on animals, but we do so out of a desire to vanquish disease and we do so within a tightly regulated world of laws designed to minimize such pain. Block has illegally and premeditatively circumvented laws, and inflicted pain on animals solely for his desire for profit. This is despicable, and deserves severe punishment, both for Block's sake and as a message to his colleagues. I strongly urge that you mete out the maximal punishment, both in terms of jail time and in terms of fines."

The Miami judge, John Kehoe, acknowledged that such responses had influenced his decision to reject the misdemeanor plea. Block then pleaded guilty to felony charges, receiving in April 1993 a 13-month jail term and $30,000 fine.

There were other scientists, though, who couldn't swallow the thought of siding with McGreal on any issue at all. Others maintained that, without question, animal importers are a vital part of the research community. That viewpoint is best illustrated by the testimony of Dr. Charles Chambers, a former executive director of the American Institute of Biological Sciences. At Block's sentencing hearing in April 1993, Chambers insisted that endangered species have no rights that should not fall before the needs of man: "If there were one orangutan left on this Earth and it was in [Zoo Atlanta] and my child needed an organ from that orangutan, I would take it."

Gormus simply has a hard time believing that Block would be so underhanded. He spent hours on the phone with Block, calling from Africa. Time after time, if Gormus didn't have the proper permits, Block refused to help transport his mangabeys out of Africa. It was frustrating, but it was by the book. He remembers Block as generous to a novice animal hunter in Africa. "He wheels and deals, maybe," Gormus says. "He has an attitude toward animals like farmers do. There are people here at Tulane who think just like him, that a monkey is a tool. I've seen them pick up a 30-pound animal and dangle it by an arm. They make fun of me because I buy bananas for my monkeys, because I like them. Monkeys to Matt are like pigs and chickens. Just animals. Whereas people, you understand, are people."

It was in Africa that Gormus began to realize that if he was too purist, if he refused to deal with wildlife traffickers, if he refused to accept a certain number of monkey deaths, he was never going to get the mangabeys at all. He was going to have to accept a certain amount of grit. That was made clear, too, in the orangutan trial. One of Block's attorneys, David Russell, asked that although Block had pleaded guilty to a felony, he still be permitted to carry a weapon. A dealer that Block had informed against was threatening to kill him. The attorney pointed out that Block already had a concealed weapon permit. Shortly thereafter—in a curious irony—Russell asked that Block be given a light sentence because "this is a nonviolent crime. It involves animals."

Gormus went to Africa with his American Express card, a clutch of traveler's checks, and several thousand in cold cash, wadded into a money belt around his waist. Today, it makes him shudder, thinking about waltzing into countries, where the incomes could run as low as $200 a year, with that fat layer of money packed around his body. The only people who robbed him, though, were the wildlife traffickers. The first animal exporter he dealt with, who will be known as John Bryant in this book, took the university for a $22,500 ride. It was one of those experiences that you replay in your head over and over: If I'd been cooler, if I'd been smarter, tougher, something, I'd still have that money. Gormus confided his unhappiness, miserably, to his diary: "I don't think anyone can blame me. I did the best I could. But still I feel like I'm to blame."

Bryant was an old hand at exporting primates. He'd built his reputation routing chimpanzees out of Sierra Leone. When Gormus met him, though, his prime interest was building a gin distillery. The two men would sit on the porch of Bryant's house, drinking African-brewed gin and talking mangabeys.

Bryant told him that because the monkeys were so hard to get, especially live ones, he would need half the money up front. One hundred and fifty monkeys, at $300 a piece, the negotiated price, came to $45,000. Gormus notified Matt Block that a big advance was required. Block wired the money to Africa. Then Gormus waited and waited for the animals. And waited and waited and flew back to Africa and flew back to Louisiana and gave up. He received one telegram, from Sierra Leone, telling him that the mangabeys were impossible to find, that the Liberians had eaten the big ones. Maybe later, Bryant promised, after the rainy season ended, he would be able to gather up some small ones. Despite several other attempts by Gormus to collect, that was the end of any promises from Bryant. It was also the clear disappearance of Tulane's $22,500. At one point his boss, Pete Gerone, contacted the university's lawyers to see if the money could be recovered by going to court. They finally decided, however, to write it off to being suckered.

Gormus did find Bryant was truthful about at least one aspect, though. Trapping live sooty mangabeys was a difficult business. Bryant had described one popular method in detail, practiced by local tribesmen. The sooty mangabey troops liked to spend the night in high groves of trees, dozing in the branches. In the rising dawn, the tribesmen would place wide nets around the trees. Then they'd set the trees on fire. The panicked monkeys would flee, blindly, down into the nets, or jump out of the trees. A few always died in the flames or from the fall. But as a monkey-catching technique, it worked every time. The other technique was simpler. In the night, hunters would go into forest with big flashlights. They'd shine the lights into the trees and they'd stare upwards. When they saw a red glimmer of eyes, reflecting light in the dark, the hunters would aim their guns between them.

One of the problems that Gormus discovered in Africa, in fact, was that no one caught live mangabeys on purpose. The mangabey hunting experts were the people skilled in killing them. Gormus kept a detailed diary in Africa. He describes in it driving miles to a village where monkey hunters were known to live. The villagers were friendly. They made him welcome and offered him food. They were baffled by his request, though. "They said they didn't know how to catch a live monkey," he wrote.

The knowledge put Gormus and Tulane in an unhappy position. They didn't want to be part of shooting mangabeys out of trees; they didn't want Tulane to be the American institution who paid Africans to slaughter monkeys. At one point during his time in Africa, Gormus came up with the idea of equipping the hunters with tranquilizer guns with night-vision scopes. They could all troop out at night, as usual, and shoot the monkeys, as usual, only using darts instead of bullets. He notified Tulane that he needed tranquilizer guns. In response, he received a furious Fax from Pete Gerone, saying that his primate center was not going to spend money to buy tranquilizer guns to pot monkeys. Further, Gerone emphasized, he was not going support any expedition that killed monkeys either. In dismay, Gormus called back home, to be told that Gerone had gone into a "temper tantrum" on this point, and that it was not negotiable.

Yet Gormus had come to believe that, if he didn't work with the African system, he'd never see a mangabey. Maybe, he thought, they should just accept that they could save some of the monkeys in a hunt. Routinely, the hunters shot the mothers. Sometimes they left the babies to die. Sometimes they would bring the youngsters home, keep them as pets until they were big enough to eat. Then they would cook them. Gormus started thinking about the monkeys that could be salvaged from the hunt. It was by that logic that he eventually started getting mangabeys out of Africa. He paid people to visit the hunting villages and buy the baby mangabeys before they went into the cook pots.

When the deal with Bryant fell through, Gormus flew back to Africa. By that time, he'd met other wildlife traffickers. He fell in with a pair of friendly bird smugglers. They were not experts in catching mangabeys, either. They were, however, willing to learn and willing to help hunt, all for a good price. Gormus was prepared to pay $300 to $400 an animal, top rates. Often the big money stays in the United States. A macaque can be bought for $5 or less in Indonesia, but it sells for $1,000 or more in the United States. Gormus had other expenses, of course. He provided his new friends with meals, transportation, housing on the road, and an unstinting supply of Guinness stout. He slept with them in mud hut villages when their old BMW broke down. He worried sometimes, as he watched them sharpen their machetes, about whether he was too obvious a target himself. He was never harmed, though. The two men went by false names. One traveled under a dead man's passport. They used several aliases; here they will be known as Barber and Cook. It was Barber who concealed from Gormus the competing dealer with the thirty-eight mangabeys in his yard. "Eventually, they split up. Barber had transported thousands of birds without paying Cook, and he found out about it," Gormus says.

Working with just Cook, the scientist started to gather mangabeys. Most of them were from the homes of hunters. Some came from restaurants. They heard about a restaurant that had three mangabeys and drove over. The kitchen had only one left, so they bought it. They heard about a man who was building a small zoo. They bought his three mangabeys.

By the end of his fourth trip, in March 1991, he'd developed a certain tolerance for his colleagues in mangabey hunting. "A lot of shady people and quirky characters," he says. "But the morality of the system is different there. Lots of times people will promise you something when they can't do it, because they're afraid of hurting your feelings [if they told you the truth]. In this culture, you would call them a liar, in that culture you wouldn't. Take Barber. He was a loudmouth and a dangerous driver, not the kind of pal I would have had by choice. But he took me into his place and introduced me to his family and friends. If I had been raised like Barber, I might have ended up being a kind of benevolent crook."

There are now 85 mangabeys at the Tulane primate center. Fifty came from Africa. Gormus also purchased 35 from the breeding colony at the Yerkes primate center. The numbers still fell short of the 150 they thought they would need for their study. They still wanted more monkeys in their study. So after all of that, Gormus added 50 rhesus macaques, purchased from China, into the leprosy studies as well. He admits the irony himself, to spend so much time hunting for sooties, to fall back on rhesus macaques after all. Mostly though, he just wanted to get the project going.

Gormus's first approach was to test a vaccine against leprosy. It was classic medicine. The monkeys were first given the vaccine, then challenged by injecting them with live mycobacterium. All the macaques were injected and 35 of the mangabeys. The rest of the mangabeys were dedicated to a new breeding colony, as required by Fish and Wildlife.

The breeding colony, in fact, may be the only result of Bobby Gormus's four trips to Africa. In the summer of 1993, NIH withdrew funding from the leprosy project, complaining that Gormus, among other things, had taken too long to start getting results. The fiasco with John Bryant is partly to blame for that. Gormus was so sure about getting the mangabeys then, he activated his three-year grant after reaching the agreement with Bryant. It took another two years to get the monkeys, meaning that he had one year's worth of work to show for a three-year project.

Considering that, he received a pretty good scientific review, scoring in the top 17 percent of all grants reviewed. He'd found some interesting hints of the relationship between the leprosy bacterium and the immune system. The little organism seemed to be able to blind the immune system in some way, so that it could slip in, unrecognized. To Gormus, that seemed remarkably like some of the strategies used by the AIDS virus. He thought the leprosy work might even yield some clues as to how infections like AIDS also dumbfound the immune system.

Still, 17 percent wasn't good enough. The funding cutoff was 14 percent. Gormus appealed to NIH. "I'm going to hang onto this by my fingernails if I have to." He argued that the program had gotten off to a slow start, but had promising results. He pointed out that an endangered species had been taken out of the wild for the work. The reply was the answer of a bureaucracy: The letter from the National Institute of Allergy and Infectious Disease, which was funding the study, acknowledged that the reviewers had recommended funding the grant, but "because of budgetary uncertainties, NIAID cannot make a commitment for support at this time."

The suspicion at Tulane is that NIH has lost interest in leprosy, which is a fierce infection in the tropics but not in this country. The hot bacterium these days is the tuberculosis bacillus and NIH, particularly the institute within it that funded Gormus's work, has been sending out requests for proposals to use rhesus macaques as a model to develop a human TB vaccine. Gormus has decided to apply for a TB grant himself. What of the mangabeys? If he can't finish the project, he's not sure. The fifty rhesus macaques are probably useless for other research because of the leprosy infection; he expects that they will simply be killed and dissected. "I hate to do it," he says. "They're really beautiful young animals." Stubborn to the end, Gormus is reapplying for NIH money, hoping to salvage the study.

The mangabeys, a protected species, cannot be killed. Instead, they will be treated with anti-leprosy drugs. Pete Gerone agreed to support the mangabey colony for at least a year, out of his center's general funds. But what then? What if a breeding colony is no longer useful for medical research? Do you continue spending NIH dollars, meant for human disease studies, to support it? The Yerkes center is studying AIDS in its mangabeys and perhaps Gormus's mangabeys will eventually go into that kind of experiment. Perhaps, Gormus thinks, the monkeys will eventually be a curiosity, a surviving remnant of the wild mangabeys of Africa.

"I don't think there's an animal species in Africa that will survive," he says. "Ghana's the country I was in that had the most plentiful supply of mangabeys. But they're cutting down the trees there to pay their debts. I saw trucks go by with trees so big they could only put three on a truck. The big logging companies pay for reforestation, but when you've got government officials who haven't gotten a paycheck for three months, where do you think that money goes? Unless AIDS kills everyone, nothing will survive there. I look at these monkeys, and I think they may be the last of their kind in the world.

"Sometimes I wonder what I'm doing, making humans live longer and longer when some of us are destroying the planet. Still, it's people. If they're sick, you just can't turn them away. And I know I'll continue doing animal research because I'm human, and we can rationalize any damn thing we do."

He won't accept, though, the charge that his project contributed to the plight of the animals. It's not that way. Every monkey he acquired from Africa was destined to be killed there. The ones he bought, shipped back to the United States, would be long gone if he hadn't purchased them for medical research. He means that as a fact of life, not as a criticism of people in Africa. "We live in a protected world here and they are closer to what's real," he says. "What really gets me is people passing judgment on this without knowing anything about it. And you can't know anything about it unless you've been there. You can't imagine what it's like for those people and those animals. I don't think the people at NIH really understand. I don't think Shirley McGreal understands. And I'll tell you what the bottom line is here. The bottom line is that I saved the lives of those mangabeys."

TWELVE

One Nation

THE FORESTS ARE also vanishing around the Yerkes Field Station, where Robert Gormus once hoped to find all his mangabeys. The station covers 117 acres of once-rural Georgia, northeast of Atlanta. Back in 1966, when it opened, the field station was surrounded by woodland; pines, maples, flowering dogwood and farmland; orchards of peach, rows of okra, tomato, yellow squash. The residents of Lawrenceville thought in agricultural terms then; the Yerkes station was "the monkey farm."

Now Lawrenceville is an upscale bedroom community for the sprawl of Atlanta. Along Collins Hill Road, the winding route toward the monkeys' home, bulldozers rumble, scraping away the stumps of pine trees, tumbling the native red clay. The turnoff for the field station remains a rough dirt road, dipping wildly, rutted by winter rains. Nearby, one glossy, new subdivision, high brick walls gleaming with brass trim, posted signs advertising 3,000-square-foot homes. The gabled rooflines of solid, colonial-style houses march above the wall in an ever-lengthening parade.

The field station sits in a hollow. It is also walled—a high chain-link fence, a triple strand of electrified barbed wire, and bright yellow "Posted" no trespassing signs clamped against the links. Behind the fence, one can see only a collection of mobile homes, part of the center's modest offices. The real money goes into the primate enclosures. You can't see them from the road, only hear their occupants—the sudden, unexpected barks and hoots of monkeys and apes.

There are some 1,600 animals at the field station, mostly rhesus macaques, but also pigtails, stumptails, crab-eating macaques, mangabeys, gibbons, and chimpanzees. All the animals live in large groups. There are 24 corrals, surrounded by high walls and fences, but open to the sky. Most hold about 80 monkeys. On foot, the station resembles a simple, monkeys-only zoo, grass and gravel paths winding from enclosure to enclosure. In rainy weather, the monkeys can retreat indoors, climbing through passageways into sheltered cages. Generally, they hang out under the sky, perched on logs and chunks of concrete pipe. School children, neighbors, government officials tour the place every year. Tom Gordon, the station's director, dismisses the wires and alarms as mostly show; "They'll keep out anyone who wants to stay out anyway."

When the subdivisions began crowding against the station's boundaries, Gordon and his boss, Frederick King, discussed moving, backing away into a more distant rural area. They decided against it, though, partly because they figured the subdivisions would follow again, sooner or later. Any possible move seemed too expensive, too upsetting to the animals, and too far from the headquarters of the Yerkes Regional Primate Research Center. The main campus is about 30 miles away, at Emory University, tucked into the hills of north Atlanta. Emory wants its scientists to be able to find the field station.

"So we work hard at being a good neighbor," Gordon says. "We thought the best way to breed resistance would be to become aloof and unfriendly. So we've done everything possible to be open and available to the neighbors. We give tours, we go to zoning battles with them, we do school programs, and in general we are very well accepted."

He has learned to be a pragmatist, working with monkeys. He is in his early fifties, fair-haired, bearded, thin, at home in blue jeans and running shoes, required wear for the tumbled landscape at the station.

Gordon graduated from the University of Delaware with a master's in psychology. Like many animal researchers—like Roger Fouts, Stuart Zola-Morgan, Duane Rumbaugh—he started out interested in the human mind and became fascinated, along the way, with the minds of other primates.

At the university, he worked summers in a lab studying stress in monkeys, putting them in restraining chairs, giving them electric shocks. The shocks were controlled by one monkey, dubbed the "executive." The boss monkey had to push a bar every few minutes. If not, all the monkeys received a mild shock. The scientists expected the monkey in power to be most stressed; it was the ethic of the time, that the person in charge, responsible for others, felt the greatest burn. The monkeys proved that wrong; the shocked and helpless animals were clearly more miserable. Now, the basic beliefs of human psychology have shifted. No one doubts that lack of power, the feeling of being trapped and helpless, is stressful. That may sound obvious today. Then, with all the emphasis on the burdens of authority, it wasn't. The researchers saw what they believed to be true.

Gordon took several lessons out of that summer job. He learned that he liked monkeys a lot. He learned that he didn't like restraining chairs. He learned that the biology of stress is complicated, too complex to be tested by strapping down a rhesus macaque and jolting it with electricity. He learned one of the biggest mistakes a scientist can make is to assume he knows the answers.

At the field station, Gordon keeps a framed color photograph on his desk. The picture is of two rhesus macaques curled around each other. The embrace seems to steam, a clasp shared by the Rhett Butler and Scarlett O'Hara of the monkey world. Furry cheeks press together; mouth is hard against mouth. Gordon smiles as he picks up the photo. "It looks romantic, doesn't it?" he asks. He waits a beat, knowing the importance of good timing. "These are actually two sisters. This was a very tense, aggressive moment. I keep it on my desk for a reason. It tells me every day: 'You don't know it all, Gordon.' "

He likes macaques well enough to get into arguments with other Yerkes staffers over whether chimpanzees are really more interesting. He admires the monkeys for their obvious intelligence and their in-your-face independence. He just plain likes to watch them. During a ramble across the station, he stops by a corral of rhesus monkeys. A female, her face flushed red, a signal that she is in prime breeding form, tries to seduce a sleepy male. She begins grooming him, the major wile of a rhesus, stroking and combing out his fur. Gently, hopefully, she teases out tangles and insects. Finally, he comes through, mounting her. But, obviously, it's a polite gesture, without real interest. He sits on top of her briefly, hops off.

"A pancake mount," Gordon calls it. The movement, though, attracts her jealous offspring, and suddenly, the couple is joined by the female's daughter and the daughter's very young son. The baby monkey is fascinated by the big male's tail. He begins to fiddle with it, tug, tug, pull. The male tries to ignore the baby, but, finally, gives a disgruntled leap away. The females gaze after him, forlorn.

Gordon stands watching them, narrating events, his voice cool and crisp, dry with humor.

It's easy to watch the frustrated female, the friendly baby, and think of people; of dates gone wrong, of infants, ever curious, poking and prying across a room. It's important, to Gordon anyway, not to do that. Rhesus aren't people; they aren't fuzzy little mimics either. "I've worked with monkeys for 25 years now, and they are still new to me on many levels. They're interesting, they're even fascinating. But they are never cutsie, and they are not like little people. I happen to think that idea is demeaning to them. If you really get to know them, you don't need to paint that on top of them."

He faults animal activists for making the monkeys too human; he faults scientists for making them too mechanical. "The degree to which you see animals not as test tubes, but as complex living beings, in a complex society, with relationships, strong family ties, friendships, alliances, the whole schmear, that's what allows you to see why they need such extraordinary care from us. If you just see animals as test tubes, I think you've made scientific and ethical errors. Science that's trivial, pointless, poorly done, without proper respect for animals, has no place in the 1990s."

It says a lot about the way primate research is changing that such a viewpoint—a challenge to the status quo—would come from inside the science community. Out of Gordon's belief in doing it right comes questions that are not only serious but speak to the future of the field. How should monkey experiments be conducted in the 1990s? Can they really take into account the intelligence of the animals, their social nature? Should they—given the precarious hold of the species in the wild—be conducted at all? If they are, should protection of the species be a required part of the laboratory program? The Yerkes Field Station makes an interesting backdrop to those questions. Not only because Gordon himself is willing to raise them. But because, to some extent, the field station represents an unusual experiment, a place where researchers have tried to balance the needs of science against the needs of the animals themselves.

The station is self-sufficient; breeding its own animals. The monkeys live in large colonies. They are trained to participate in blood tests or to come into a small cage for an injection. It takes time and patience; monkeys are not domesticated or even anxious to please, as a dog might be. You cannot train them to come running when you call. With patience, with food rewards, they will learn to move

into a cage, hold out their arms for a blood draw. In other laboratories, monkeys have been trained to walk over to a restraining chair
and sit down. At the field station, the staff has taught monkeys to
move out of open corrals, into tunnels that lead to smaller cages.
Once there, the animals will extend their arms for blood to be taken,
turn their bodies to receive an injection.

Using that technique, Gordon and his colleagues have made some
forays into the body chemistry of masculinity, among other things.
They once tracked testosterone levels in male rhesus macaques for
some three years, taking blood samples monthly. Every time the
male hormone went up, so did both sexual activity and increased
aggression. Curiously, testosterone was also a measure of success.
The hormone rocketed when a male became dominant in his
group—the king of the hill, so to speak—and tumbled when he took
a beating from other males.

Gordon admits that the approach is more complicated, more difficult and more time-consuming than parking a monkey in a small
box, sticking a needle in, and drawing out blood. In the intense
study of an infectious disease, an experiment requiring a surgical
procedure, you couldn't use it. On the other hand, you couldn't get
those testosterone results—watching the hormone drive aggression
against others—in an isolated monkey. What would sexual dominance mean in a monkey that never even touched another? "These
animals didn't evolve to live alone, remember," Gordon says. "We
see this as a bridge between nature and the lab environment. We
develop and try to promulgate this approach. I think our central
achievement at this station is that we have socially housed animals
and yet we have proved that they are accessible for experiments,
that you can have both."

To their credit, animal advocates have made it very hard for biologists to test animals the way geologists might break apart rocks or
meteorologists dissect the chemistry of a cloud. As Duane Rumbaugh points out, we have come away from the time of Descartes
and his belief that animals are merely walking machines. There are
recent lessons, reminding us of how far we have traveled, sometimes
painfully. The still-dramatic tapes from Thomas Genarelli and his
baboon head-injury experiments are not easily forgettable. Neither
is the anger they provoked. One of the paradoxes, though, is that
the scientists themselves have done much of the work that gives
animal activists so much power.

The kind of work done by Rumbaugh and others provides a

framework for protest. It allowed activists to insist Genarelli's baboons were not animated rocks. That the baboons were smart animals trapped in a desperate situation. Every demonstration of animal intelligence emphasizes the point; we are not experimenting on unthinking creatures. Without intending to, scientists and activists have worked together on this point—one side providing the information, the other spreading it.

When you think about Duane Rumbaugh's macaques, competitively shooting down computer targets, trying to outdo each other, can you think of them as animals too stupid to deserve respect? At the Yerkes Field Station, scientists have analyzed the calls and chatters of macaques, working out the patterns of the sound waves. They've found a language, unheard by the human ear. Baby rhesus, for instance, when they squeak for their mothers have different squeaks. One chirp will announce that a relative is bothering me; another declares that a stranger is scaring me. The Yerkes scientists have found that on the first squeak, the mothers barely look up. On the second, the mothers bolt across the yard, bristling, teeth bared. Can you think of the rhesus, once you know that, as animals too simpleminded to communicate, too primitive to have feelings?

Even the work of Harry Harlow, for all its dark and haunting aspects, emphasized graphically that rhesus macaques were animals that did not, really, survive alone. Who could see a baby monkey, huddled on the floor of a cage, and not wish for it to be returned to its mother? Harlow's work, in some ways, gives away the sheer complexity of the issue—that his work could be so painful in many ways, and yet also influential in forcing people to recognize the needs of captive primates.

"What's really fascinating about Wisconsin and Harry Harlow is that Harlow, in one way, did terrible research," says Frans de Waal, who worked at the Wisconsin center for ten years before coming to Yerkes in the early 1990s. "On the one hand, he deprived them of all love and affection. On the other, that produced so much of an effect that people had to talk about it. The Wisconsin people got into primate welfare early, maybe because of it. Look at the people who came out of Harlow's program. Or look at Viktor Reinhardt today. It's a very strange history."

De Waal's specialty is animal behavior. He began studying primate intelligence in his native Holland, made his reputation working with chimpanzees. It was his work that made people realize that, like humans, chimpanzees not only feud in anger, but do it intelligently, plotting battles. And also, like humans, they engage in diplomacy, soothing each others' feelings, making peace offerings to end fights. De Waal has taken that issue even to the notoriously unpeaceable rhesus macaques. In one recent experiment, he raised young rhesus with a closely related species, the more laid-back stumptail tribe. To his fascination, the rhesus quickly picked up the habits of the stump-tails, easing themselves out of arguments. Somewhere in that combative nature, obviously, was also an appreciation for peace.

Rhesus macaques—or any other primates—cannot just be dismissed as simple creatures. Their abilities go far beyond the basic skills of food-finding and nest-building. These are animals that teach each other negotiating skills, learn to operate computers, recognize their kinfolk from a photograph. They are intelligent, capable, quick learners. They are, like us, complex beings. Once we recognize that, we must also recognize that the choices we make in using them are complex, too. It might once have been easy to toss a monkey into a research project, taking no particular thought. Today, the reverse is true. We should hesitate and we should think.

If you accept, at some level, animals as necessary in biomedical research, then the point is that you use them wisely, trying to understand them. You work with them. You don't treat them cheaply. "I think that if you start by taking a chimp heart to replace a human heart, we may be on the wrong track," de Waal says. "One on one—one chimp or one baboon for one human, I'm not sure that's a good use of an animal. On the other hand, with something like AIDS, there are 40 million people infected. If you could stop that, with the use of 1,000 chimpanzees, who I am to say we cannot? Sometimes, these are presented as very simple issues. And that is not honest."

That point of view, again from within the science community, emphasizes how incredibly influential the last 15 years have been in raising awareness. The animal advocacy movement has changed the way all of us—in science and out—think about the use of animals. That's not to say that most Americans don't support animal research. The most objective polls suggest that they do. People do not, however, support it without qualification, without question.

There is growing criticism of the so-called "frivolous" uses of animals in research, such as testing new cosmetics in the eyes of rabbits. Monkeys are not used in cosmetic testing; in that field, rodents are the animal of choice. Yet, such concerns touch all aspects of animal research. There's a ripple effect. If society becomes dismayed by the suffering of rats and rabbits, that compassionate reaction inevitably spreads to other species. The concerns are more than talk. Under pressure, with new social awareness, many cosmetic companies are now pulling out of animal testing, seeking alternative research methods.

In biomedicine, too, there is no longer an unquestioned, openhanded use of animals. You could not, today, kill some 2 to 5 million monkeys in AIDS research, the way you could in polio work four decades ago. Look at the pigtail macaque model for AIDS; Seattle researchers have been criticized, not only by activists but by other scientists, for importing a mere 1,000 extra monkeys, much less hundreds of thousands.

You could not today get federal approval for some of Harry Harlow's studies, taxpayer support for building "pits of despair" in which to drop baby rhesus macaques. Even the most ardent supporters of Taub's research with the Silver Spring monkeys say his experiments might not be approved today; the standards of his laboratory are those of 15 years ago, not of now. Back in the days of the polio race, of course, the general public didn't really question what scientists did. Neither did they doubt Harlow in his prime. It used to be that scientists were untouchable. In their own opinion, and people generally agreed, researchers were doing work too important, too brilliant, too difficult for the rest of us to understand, much less criticize.

Yet, science does not stand somehow above the basic laws of decency and respect for life. No one has to be a fully trained psychologist to recognize the suffering entailed in maternal separation experiments, like those Harlow conducted. Those of us outside the realm of science have a right, and perhaps an obligation, to hold researchers to a standard of compassion. In recent years, people have been less blindly admiring of science, more willing to hold it to the moral obligations of everyday living. If that means we are all less awestruck by science, that means we are moving in the right direction.

For many scientists, that transition into the public arena has been unwanted and unwelcome. It's not surprising. Who wants to go from unconditional approval to wary scrutiny? Yet, in the end, this is clearly a necessary transition. The rest of us—those who don't spend our days in laboratories—deserve to know what happens inside those walls. And we should have some influence. It's not just that the money comes out of our pockets—some $6 billion a year for NIH alone—but that the research has changed and continues to change our lives. The flashy success of the polio work, the disappointing obstacles of the fight against AIDS, all of that matters to us, to the quality of our lives and to our survival.

In the last decade of the twentieth century, it is clear that the future will continue to be shaped by science and technology. Perhaps more so than ever before. To function in that world, people need to understand the forces changing their lives. That means scientists must learn that their job description has changed. They do not work only in a laboratory, they work for and with the rest of us, with all the risk that entails. They must realize that the new job description may include controversy and anger. Some will probably need to learn, as Stuart Zola-Morgan and Seymour Levine have, that nasty letters and even public embarrassment are survivable. They must learn, further, that if science is worth explaining and defending at all, it must be worth explaining and defending not just to those who already approve, but to a broader constituency. If the price is controversy, the reward is understanding—a gain in public knowledge and a respect grounded in reality instead of some idealized image of Dr. Genius, mixing miracles in test tubes.

On the other hand, that openness is not going to come if animal advocates don't move away from the gates, at least a little. Whether or not they agree, advocates need to give scientists a fair chance to describe the work honestly, without subterfuge, without what Tom Gordon calls "defensive press tactics." Break-ins at laboratories and hostile vigils outside researchers' homes have the unfortunate result of doing just the opposite. The activists' most aggressive strategies have driven animal research behind barricades. They have made researchers run for cover, helping to make science less public, less open, and in the end less well understood. It is time for activists to realize that guerilla tactics only get in the way of good intentions and good ideas.

Gordon, at the field station, sees it as a time of evolution, tough to live through, but moving toward something ultimately better. "Ministers and priests were once put on pedestals. Now they're being sued. I would argue partly, and I wouldn't get the agreement of every one here at Yerkes, that it's okay for scientists to be off the pedestal, too, for us to be questioned about what we do. Some scientists, especially senior ones, appear to have a difficult time making the transition from having everything our way, from not being accountable, to being fully accountable. Personally, I think that's an advance."

It is too simplistic for researchers to sell science, as they some-times do, as a one-dimensional tradeoff; the life of an ugly rat or mouse or monkey for that of a beautiful child. It is too simplistic for animal rights people to define the research community as a bunch of would-be butchers on the loose, sharpening their meat cleavers in the backyard. It's not just that the issues are complex, the science is complex, too. Consider the delicate neuroscience of Stuart Zola-Morgan, threading his way through the secret realms of the brain. To dismiss it only as the work of Dr. Zola-Morbid, monkey killer, as California activists have done, misses the point. In all the discus-sion of educating the public, the emphasis seems to be more on pro-paganda, the portrayal of the happy research monkey versus the portrayal of the suffering lab animal, and nothing beyond. If taught on that cartoon-like level, how can anyone hope to decipher the complexities of AIDS research, understand why the answers are so slow in coming? How can anyone follow the progress of Zola-Morgan and his colleagues? The rest of us deserve a better education than programs selling science and programs attacking it. It is, after all, our future that we seek to understand.

One issue, debated by both sides, is that of "alternatives" to ani-mals in research, such as the use of computer models, tissue and cell cultures, increasingly sophisticated chemical tests. So far, the em-phasis has not been on replacing monkeys but on animals used for less complex work. Cosmetic testing is an example. And many of the replacement technologies have focused on toxicity testing, such as using cell culture studies to replace live rabbits and rats. There is the promise, of course, that more sophisticated alternatives may come. Scientists are exploring the idea of studying nervous-system reactions in small groups of cells; of developing a line of heart cells that would individually beat, for use in studying the heart. Re-searchers are gaining confidence in computer models, so they can study connections within the body or track the genetics of disease.

If the use of alternatives is progressing slowly, it has undoubtedly been held back, partly, by sheer bad attitude. Animal activists keep trying to shove scientists into total reliance on alternatives. Re-searchers resent it. It is not true yet—no matter how many animal advocates wish it—that technology can replace all animals in biologi-cal research. A computer program cannot find the answers to AIDS when no one even knows what questions to ask. And the AIDS virus has been just as frustrating in the test tube as in living bodies; many treatments that look great in glass collapse when tried in an animal. On the other hand, the field could use greater support from scientists in general. Perversity—resisting change just because activists support it—is not a good reason to turn away from promising technologies.

In the ideal approach, animal advocates and researchers could bring their viewpoints together in education programs, balancing the issues. The goal is not necessarily to achieve comfort but, with luck, a realistic understanding. People need to see animal research clearly, with its costs in life and blood, with its benefits in knowledge and medical treatment. Such a vision may bring us no farther than a place sometimes called "the troubled middle." Yet, there, we may accept complexity. We may learn there is no requirement to like every aspect of the choices we make. Some tension is an inevitable part of every thoughtful decision.

There should be room for a Roger Fouts to argue for chimpanzees, just because he thinks they are wonderful, without losing the respect of his colleagues. Scientists should be able also to disagree openly with animal advocates, without fear of retribution. Yerkes, for instance, has a tough and antagonistic relationship with the animal advocacy movement. It's outgoing director, Frederick King, has been harshly critical of animal activists. In turn, the activists have staged annual, angry protests at Yerkes, describing it as research hell. At laboratories such as LEMSIP, where Moor-Jankowski has a friendly and open relationship with activists, such protests simply don't occur.

Yet, like LEMSIP, Yerkes runs a bright and friendly nursery for its young chimpanzees, plenty of cuddling by humans, plenty of toys. And LEMSIP has nothing like the Yerkes field station or even the indoor-outdoor chimp runs at the main Yerkes headquarters. That's partly due to a difference in climate; the icier New York winters limit housing primates outside. It's also due to Moor-Jankowski's fear of diseases. The animals stay caged indoors, sheltered from wild viruses. There are no big, chatty social corrals, as at the Yerkes field station.

Animal advocates, though, do not rally for better housing conditions at LEMSIP. They have a good relationship with Moor-Jankowski; he has kept his facility open to them. "Why can't they all be like Moor-Jankowski?" asks Alex Pacheco of PETA. And Moor-Jankowski earned the goodwill with his careful management of his facility and his straightforward relationship with would-be critics. Yet Yerkes deserves goodwill, too, at least for its enlightened social housing programs, its efforts to work with the monkeys.

Neither does the Yerkes Field Station stand alone. More and more scientists are painstakingly training their animals. Chimpanzee researchers in Texas have taught their animals to pee in a cup, meaning that they don't have to anesthetize the chimps to get a urine sample. At the California primate center, monkeys have been trained to put their arms into blood-pressure cuffs. Other primate research centers have pushed hard, as well, to bring monkeys into groups. The National Institutes of Health, under Steve Suomi's direction, has created an experimental colony of 40 rhesus macaques in a five-acre enclosure. They are monitored by telemetry, another technological advance that will allow scientists to watch monkeys without constantly tinkering with them. "It's an evolving field," Gordon says. "What was acceptable 10 years ago is probably marginal, at best, now. What was marginal 20 years ago, is probably no longer acceptable at all."

When PETA acknowledges such efforts, when scientists can accept the legitimate concerns of activists, then perhaps the rest of us will have to endure less name-calling and mudslinging. Perhaps the basic issue—the balancing of human and animal needs—can be discussed in the depth it deserves. And when both sides are putting less energy into dislike, more into cooperation, the monkeys also will benefit. The issue of saving primates in the wild is a perfect example of that.

If you talk to Shirley McGreal and Frederick King separately about conservation, they can sound surprisingly alike—worried about monkeys, worried about the future. "One of the reasons I don't think the animal movement is going to go away is that the issues are so fundamental," King says. "I'm talking about the balance of nature in the world, the relationship of organisms to other organisms. I think we may be beyond the point of saving our world. I think we may have already gone too far. It's a horrifying thought. Yet, we're always 50 to 100 years behind in our knowledge, and when we destroy the rainforests, take away the planet's oxygen source, we can destroy the globe we live on. Many animal rightists are ardent conservationists and preservationists, and I would certainly join any group in that."

Yet, still many scientists insist on their absolute right to any monkey in the jungle. They unite against efforts by McGreal or a recent campaign by the Humane Society of the United States to ban the importation of wild primates, to insist that researchers rely on captive-bred animals. They denounce such moves as plots, to drive up their costs and harm their profession. There's no doubt that limiting access to wild monkeys will limit some research and make monkeys more expensive. Yet, realistically, so will running out of wild monkeys. We should limit our use of monkeys for ethical reasons, not the least being our close genetic relationship. We should also limit for practical reasons. Monkeys are also a vanishing resource. Use them without intelligent protection and you will use them up.

The plight of primates is so dire that even Charles River, the largest primate importer in the country, has made plans to phase itself out of the wild monkey business. In a letter to the Humane Society of the United States, company president James Foster said Charles River hoped to be out of the rainforests by the end of 1994. "We have planned for many years to 'phase out' the importation of feral primates once a sufficient quantity of purpose-bred animals were produced to meet the essential research needs of our customers," Foster wrote. The company plans instead to sell captive-bred macaques, raising them in colonies in the United States and Indonesia.

In a world of limits, biomedical researchers cannot be exempted from trying to conserve, trying to use wisely. The question of whether we should continue to trap wild animals for research should not be an antagonistic one. Scientists who consider it merely a ploy by the enemies of science are reacting blindly. Given the good performance of breeding colonies here, why shouldn't we begin phasing out the use of wild animals? There are some good scientific reasons for doing so. No one goes out and traps wharf rats for a rat study, grabbing them out of whatever sewer, carrying whatever disease. Researchers want to know the full history of each animal, its family tree, its diseases. Yet a wild monkey—out of some forest, carrying some selection of diseases—seems to be totally acceptable. It's a curious double standard and, probably, an obsolete one.

As tropical countries move to increase their own protection of vanishing species, there are obvious ways for this country's scientists to support that. They can contribute to breeding programs and conservation studies abroad. The federal primate centers have moved in that direction: Yerkes has a program in Belize; the Wisconsin center does work in Thailand; the California center has sent teams to Panama, to try to help ailing howler monkeys. If one looks toward the future, one question for the primate centers is whether they should become more involved in such efforts. Another is how to balance the needs of individual monkeys—with changes such as social housing—against the protection of the species.

The fact is, progress in caring for monkeys cannot be judged solely by numbers. Attention to the needs of individual monkeys may come at some cost to the population. Primate centers are going to lose some monkeys if they pay more attention to the "psychological well-being of primates." If they increase social housing, more are going to be injured, more of them are going to get sick. The whole business of isolating monkeys came about, in part, because in the early days, monkeys were tossed in cages together and they became ill in droves. Moving away from that is a tradeoff. Even the strongest supporters of social housing, such as University of Georgia psychologist Irwin Bernstein, acknowledge the risk.

"If I were sitting on an animal care and use committee, it would take me a long time and a lot of argument before I would ever agree to let you routinely house animals in isolation, long term, regardless of what experiment you were doing. Think about a human being hit by car," Bernstein says. "What happens if you put him in a room, hooked up to tubes, and never let him see his family. This guy is going to be certain he's dying. He's deprived of every social contact, everything he needs. So here's an animal living someplace, trapped, captured, handled, prodded, poked and manipulated, now dumped into a little cage with strange food in a strange building. At least if he had a social companion, that would be something to hang onto. A lot of animals have trouble coming through quarantine. They die there. Maybe it's the stress. Maybe it's time that we weigh the risks of spreading disease against the risks of killing the animal through isolation. On that point, I don't think that humans and monkeys are two different nations."

But as it stands, the research community and its activist critics are like two different nations, nations locked in a long, bitter, seemingly intractable political standoff, weapons at the ready. They are fighting the monkey wars. "There's a lot of negativity," de Waal says. "If you talk to people both on the animal rights side and the research community, what you get is the feeling that people hate each other. It's very polarized, no positive feelings either way. I like to imagine what is the primate center I would like to see in my lifetime. I think, to get there, we have to get rid of all this hate."

Voices like his, like those of Tom Gordon, Duane Rumbaugh, Jeff Roberts, Jonathan Allan—calm and rational in the middle of anger—are a mark of progress. Places such as the Yerkes Field Station—with its open housing and lively groups of monkeys—also mark progress. But there will be too little opportunity to build on what has already been accomplished unless the animal advocates open their eyes and acknowledge such progress, and unless more scientists take a deep breath and learn to do the same. If you listen hard, there really are people on both sides willing to accept and work within the complex middle. When they can be freely heard, then we will have progressed to another place, beyond this time of hostilities.

If we—those who love animals, those who love knowledge, those who care about the future of the earth and its creatures, including the human variety—if all of us are going to find common ground on these divisive issues, then perhaps the best starting point is the one that Roger Fouts suggests.

Begin with chimpanzees, our genetic next-of-kin. So much like us, they are easy to respect. From there it is only a short path to other apes and then to monkeys. With new revelations about primate abilities, the gaps seem to be narrowing all the time. If we can come to agreement on how to regard and care for those extraordinary animals, that will mark a hard-won and much-needed lesson in stewardship. It may, with luck, carry over as we try to resolve our differences on the care of other animals, of the world at large. In working together, we may learn that we have been warring too long, and that we still are, though we often forget, one nation.

NOTES

Chapter 1. The Outsider

This book is based primarily on personal interviews. Those interviewed for this chapter include Roger Fouts and Deborah Fouts, at Central Washington University; Duane Rumbaugh, at Georgia State University; Christine Stevens, at the Animal Welfare Institute; Milton April, at National Institutes of Health; Frans de Waal, at Yerkes Regional Primate Research Center; Roy Henrickson, University of California, Berkeley; and Shirley McGreal, at the International Primate Protection League.

The genetic variation between humans and chimpanzees is discussed in Jared Diamond's book *The Third Chimpanzee* (HarperCollins, 1992). Diamond also discusses the endangered species issue, as do many other authors, including Jane Goodall and Dale Peterson in *Visions of Caliban* (Houghton Mifflin, 1993).

Mirror recognition in chimpanzees was reported by George Gallup, Jr. in several articles in *Science* in 1970 including "Chimpanzees' Self-Recognition". See January, February, and March issues. The topic is also discussed in *How Monkeys See the World,* by Dorothy L. Cheney and Robert L. Seyfarth (University of Chicago Press, 1990).

References for the section on the history of chimpanzee language research, as well as Fouts's personal recollections, include *The Ape People,* by Geoffrey Bourne (Signet Books, 1971); *The Monkey Puzzle,* by John Gribbin and Jeremy Cherfas (Pantheon Books, 1982); *Silent Partners,* by Eugene Linden (Times Books, 1986).

The early training of Viki and Gua is also well discussed in "Language Comprehension in Ape and Child," a monograph of the Society for Research in Child Development, authors Sue Savage-Rumbaugh, Jeannine Murphy, Rose Sevcik, Karen Brakke, Shelly Williams, and Duane Rumbaugh. The same monograph provided background for the Rumbaugh's work. The work with Lana is also featured in Gribbin and Cherfas's *The Monkey Puzzle,* including a description of the "cabbage" exchange between Lana and her trainer. Rumbaugh talks about his career in animal intelligence work in "The Undaunted Psychologist: Adventures in Research," edited by Gary Brannigan and Matthew Merrens (Temple University Press, 1992), pp. 90–109.

Details of the habits of Fouts's five chimpanzees come primarily from quarterly publications of "Friends of Washoe," from 1985 to 1991. Fouts writes both about his first encounter with Washoe and the story of how he thought he had allowed Washoe to poison herself in the Fall 1985 issue.

Concerning Fouts's inability to get funding, the comments from the NSF review panel are taken directly from a copy of an agency Evaluation of Proposal, NSF Form 1, April 1984.

The section on animal activism begins with the story of SEMA, Inc. That story has been widely publicized and information about it is detailed in newsletters from groups such as the International Primate Protection League, as well as in newspapers. In her latest book, *Visions of Caliban,* co-written with Dale Peterson (Houghton Mifflin, 1993), Jane Goodall devotes part of a chapter to the SEMA story.

The controversy over splitting the endangered species listing is set forth in a series of documents. They include "Comments Concerning Proposed Rule, Department of Interior Endangered and Threatened Wildlife and Plants: Proposed Endangered Status for Chimpanzee and Pygmy Chimpanzee," Federal Register Notice 54 FR 8152, April 25, 1989; letter from Robert Whitney to the Hon. Gerry E. Studds, September 15, 1989; Research Resources Reporter, "Successful Breeding Program Benefits Chimpanzee Conservation and Research," published by NIH; letter from Robert Whitney to the U.S. Fish and Wildlife Service, April 28, 1992.

Two legal documents are cited in the section on the challenge to the Animal Welfare Act. The first is the initial lawsuit, filed in U.S. District Court in Washington, D.C., in 1991. In addition to Roger Fouts, the plaintiffs in that lawsuit are the Animal Legal Defense Fund; the Society for Animal Protection Legislation, Inc.; Dr. Bernard Migler, individually and as president of Primate Pole Housing; and William S. Strauss, an attorney, a member of the Animal Legal Defense Fund, and a member of the Institutional Animal Care and Use Committee at the State University of New York, Brooklyn. The other document is the formal opinion of Judge Richey, filed February 25, 1993.

Chapter 2. Of Street Toughs and Target Practice

Interviews for this chapter included Roger Fouts; Duane Rumbaugh; Roy Henrickson; Jan Moor-Jankowski, at LEMSIP; Peter Rapp, at the Salk Institute for Biological Studies in La Jolla, California; and Rodney Ballard, at NASA-Ames in Mountain View, California. Ballard died in 1993 following a brief illness.

References for the work of Duane Rumbaugh and David Washburn with the two rhesus macaques Abel and Baker include "Perceived Control in Rhesus Monkeys: Enhanced Video Task Performance," by David Washburn, William Hopkins, and Duane Rumbaugh, *Journal of Experimental Psychology,* Vol. 17, No. 2, 1991; "Rhesus Monkeys, Video Tasks, and Implications for Stimulus-Response Spatial Continuity," by Duane Rumbaugh, Kirk Richardson, and David Washburn, *Journal of Comparative Psychology,* Vol. 103, No.1, 1989; "Ordinal Judgments of Numerical Symbols by Macaques," by David A. Washburn and Duane Rumbaugh, Psychological Science Research Report, May 1991; "Testing Primates with Joystick-Based Automated Apparatus: Lessons from the Language Research Center's Computerized Test System," by David A. Washburn and Duane M. Rumbaugh, *Behavioral Research Methods: Instruments and Computers*, Vol. 24, No. 2, 1992.

References for the history of primate research include "The Deluge and the Ark," by Dale Peterson, (Avon Books, 1991), which details the use of monkeys in the late nineteenth century and gives examples of painful experiments in this century; *Of Mice, Models and Men*, by Andrew Rowan (State University of New York Press, 1984), which in chapter 4 provides historical notes on animal research and anti-vivisection, and in chapter 8 provides an analytical look at the development of the polio vaccine; *Men and Apes*, by Ramona and Desmond Morris (McGraw-Hill, 1966), which provides a detailed look at the use of monkeys in the nineteenth and early twentieth century. Background on Sernalyn can be found in *The Encyclopedia of Psychoactive Drugs*, and *PCP: The Dangerous Angel*, by Marilyn Carroll (Chelsea House, 1985).

For numbers of animals used in research, see the Fiscal Year 1991 edition of Animal Welfare Enforcement, the Report of the Secretary of Agriculture to the President of the Senate and the Speaker of the House of Representatives. For background on polio, see Rowan's book, *Of Mice, Models and Men*; Monkeys and Apes," by Gotthart Berger (Arco Publishing, 1985); and "FDA Regulatory Use of Primates," by James H. Vickers, in *Lab Animal*, April 1983. For background on thalidomide, see the news stories: "Drug of Infamy Makes a Comeback," by Sally Squires, *The Washington Post*, July 9, 1989, p. 18; and "Still on Guard, Frances Kelsey, the FDA Scientist Who Halted Thalidomide Is Still in Fray," by Amy Linn, *The Philadelphia Inquirer*, May 8, 1988.

The listing of the use of monkeys in medical research breakthroughs is derived from a list of "Biomedical Advances Made Possible by Animal Research," distributed by the National Association for Biomedical Research. A detailed summary of the use of primates in medicine is "Primates," by Frederick A. King, Cathy J. Yarborough, Daniel C. Anderson, Thomas P. Gordon and Kenneth C. Gould, *Science*, June 10, 1988. The discussion of vision and eye work in primates comes largely from this publication. Information on the regional primate centers comes from "Regional Primate Research Centers," an NIH publication, July 1992.

Chapter 3. The Black Box

This chapter is primarily based on interviews with Stuart Zola-Morgan, at the University of California, San Diego; David Amaral, at the State University of New York, Stony Brook; and Larry Squire, at the University of California, San Diego. There were also background discussions of the work with representatives from animal advocacy groups, including Alex Pacheco of PETA and Shirley McGreal of the International Primate Protection League.

The references on the neuroscience work are highly technical, for the most part. A listing of the ones most accessible and most essential to this chapter follows:

Stuart Zola-Morgan and Larry Squire, "The Neuropsychology of Memory," reprinted from *The Development and Neural Bases of Higher Cognitive Function*, Vol. 608 of the Annals of the New York Academy of Sciences, December 31, 1990

L.R. Squire, B. Knowlton and G. Musen, "The Structure and Organization of Memory," *Annual Review of Psychology*, 1993.

Larry R. Squire, "Memory and the Hippocampus: A Synthesis from Findings with Rats, Monkeys and Humans," *Psychology Review*, Vol. 99, No. 2, 1992.

Larry R. Squire and Stuart Zola-Morgan, "The Medial Temporal Lobe Memory System," *Science*, September 20, 1991, which was one of the major announcements of the findings that the cortex was involved in memory function.

Stuart Zola-Morgan, Larry R. Squire, David G. Amaral, and Wendy A. Suzuki, "Lesions of Perirhinal and Parahippocampal Cortex That Spare the Amygdala and Hippocampal Formation Produce Severe Memory Impairment," *The Journal of Neuroscience*, December 1988.

Pablo Alvarez-Royo, Robert P. Clower, Stuart Zola-Morgan, and Larry R. Squire, "Stereotaxic Lesions of the Hippocampus in Monkeys: Determination of Surgical Coordinates and Analysis of Lesions Using Magnetic Resonance Imaging," *Journal of Neuroscience Methods*, Vol. 38, 1991, which details the variation in the brains of crab-eating macaques.

Stuart M. Zola-Morgan and Larry Squire, "The Primate Hippocampal Formation: Evidence for a Time-Limited Role in Memory Storage," *Science*, Octber 12, 1990

Philip J. Hilts, "A Brain Unit Seen as Index for Recalling Memories," *The New York Times*, Octber 1, 1991, which was the major popular press announcement of the San Diego team's new discoveries in memory function.

The information on the memory in sea slugs comes from "Biological Memory," by Larry R. Squire and Alberto Oliverio, in *Birth and the Frontiers of Neuroscience*. The discussion of rat memory tests comes from "Individual Differences in the Cognitive and Neurobiological Consequences of Normal Aging," by Peter Rapp and David Amaral, reprinted from *Trends in Neuroscience*, Vol. 15, No. 9, September 1992.

Sources for the comparison of human and monkey brains include "Empathy and Brain Evolution," a brief discussion of Herbert Terrace's work, *Science*, February 26, 1993, pp. 1250–1251. The data on encephalization coefficients can be found in "Escape from Stupidworld," by Christopher Wills, *Discover*, August 1991.

John Gribbin and Jeremy Cherfas discuss the rate of brain growth in humans versus monkeys in *The Monkey Puzzle* (Pantheon Books, 1982). Another excellent reference on the comparison between human and nonhuman primate brains is *The Lopsided Ape*, by Michael C. Corballis (Oxford University Press, 1991).

Chapter 4. The Trap

Interviews for this chapter, representing scientists, included Seymour Levine, at Stanford University; William Mason, at the University of California, Davis; Steve Suomi, at the National Institutes of Health; Gene Sackett, at the Washington Regional Primate Research Center; and Helen LeRoy, at the Harlow Primate Laboratory at the University of Wisconsin, Madison. Also from the animal activism viewpoint, Suzanne Roy, from In Defense of Animals, and Martin Stephens, of the Humane Society of the United States.

The sections from letters to Seymour Levine are taken from photocopies of correspondence.

Examples of past experiments on animals come largely from the information reports of the Animal Welfare Institute, in particular the January–February 1962 and March–April 1962 editions. Other case sources include *The Deluge and the Ark*, by Dale Peterson (Avon Books, 1991), in particular the chapter "Cages" on primates in captivity.

The history of Harlow's work is widely documented. Sources include *From Learning to Love: The Selected Papers of H.F. Harlow*, edited by Clara Mears Harlow (Praeger, 1986) and *The Human Model: Primate Perspectives*, by Harry F. Harlow and Clara Mears (Wiley, 1979).

Other references include *Mother-Infant Separation in Monkeys*, by Billy Mack Seay, a Master's Thesis at the University of Wisconsin, 1962, and *Experimental Production of Depressive Behavior in Young Monkeys*, by Stephen John Suomi, a Doctoral Thesis at the University of Wisconsin, 1971.

The story of the female orangutan is from an address given by Harry Harlow at the University Club, Madison, Wisconsin, in 1959, titled "My Life with Men and Monkeys." Harlow also describes the building of his laboratory and his growing interest in rats, over monkeys, in that address, as well as the two previously noted books.

Possibly the best overview of Harlow's attitude toward women is in a *Psychology Today* article, April 1973, called "Harry, You Are Going To Go Down in History as the Father of the Cloth Mother," a question-and-answer session with psychologist Carol Tavris. Tavris recounts that when she toured Harlow's laboratory, and expressed dismay at the sight of the cowering, isolated baby monkeys, Harlow told her that her sympathy response was a natural result of her femininity. "I'd be worried about you if you didn't think they were cute," he said.

A former student of Harlow's, John Gluck, suggested that Harlow was imprisoned in a kind of academic loneliness at the end of his career. That comment comes in a tribute to Harlow, published by the *American Journal of Primatology*, Vol. 7, 1984. Gluck's is titled "Harry Harlow: Lessons on Explanations, Ideas and Mentorship. A similar tribute was published by *Developmental Psychobiology*, Vol. 20, No. 5, 1987, by Leonard Rosenblum, titled "Harry F. Harlow: Remembrance of a Pioneer in Developmental Psychobiology." Rosenblum recounts his mishap with the artificial monkey head in that publication.

Martin Stephen's look at the experiments is called "Maternal Deprivation: Experiments in Psychology," a critique of animal models. It was published in 1986 for the American AntiVivisection Society, the National Anti-vivisection Society, and the New England Anti-vivisection Society.

Sackett's response was published in *Contemporary Psychology*, Vol. 33, No. 1, 1988. It is titled "Animal Rights, Human Rights, Scientific Rights: Who's Right?" Further information on Sackett's work can be found in "A Nonhuman Primate Model for Studying Causes and Effects of Poor Pregnancy Outcomes," *Preterm Birth and Psychological Development* (Academic Press, 1981). He also wrote a chapter on "The Human Model of Psychological Well-being in Primates," in *Through the Looking Glass*, edited by Melinda A. Novak and Andrew J. Petto (American Psychological Association, 1991).

Bill Mason has written thoughtfully about the ethics of primate research for many years. Two outstanding examples are "Minding, Meddling and Muddling Through," *Laboratory Primate Newsletter*, January 1979, and "Premises, Promises and Problems of Primatology," *American Journal of Primatology*, Vol. 22, 1990, pp. 123–138. He writes about the social needs of monkeys in "Effects of Social Interaction on Well-Being: Development Aspects," *Laboratory Animal Science*, August 1991. His work on the titi monkeys has been widely published; one of the best examples is "Parental Division of Labor and Differentiation of Attachments in a Monogamous Primate," by Sally P. Mendoza and William A. Mason, *Animal Behavior*, Vol. 34, 1986.

Seymour Levine provided me with two in-press papers on his current work, one a review article, "The Psychoendocrinology of Stress," by Seymour Levine; the other, reporting on some new findings he has made with rats (which he still continues to work with), titled "Maternal Regulation of the Hypothalamic-Pituitary-Adrenal Axis in the Infant Rat: The Roles of Feeding and Stroking," by Deborah Suchecki, Patricia Rosenfeld, and Seymour Levine.

Chapter 5. The Face of Evil

Interviews for this chapter include Marion Ratterree, Jim Blanchard, Peter Gerone, all from the Tulane Regional Primate Research Center; Adrian

Morrison, National Institutes of Health and University of Pennsylvania; Alex Pacheco, at PETA; Christine Stevens, at the Animal Welfare Institute; Susan Lederer, at the Pennsylvania State University; Shirley McGreal of IPPL; and Roy Henrickson, University of California at Berkeley.

The history of the Silver Spring monkeys comes from a wide range of sources as well as interviews. The ones I used include "The Strange Ordeal of the Silver Spring Monkeys," by Peter Carlson, *The Washington Post Magazine*, February 26, 1991; "The Raid at Silver Spring," by Carolyn Fraser, *The New Yorker*, April 19, 1993; "Report of the Subcommittee to Investigate the Case of Dr. Edward Taub," *Neuroscience Newsletter*, Vol. 13, No. 2, March 1982; Letters to the Editor, January–February 1983, *Neuroscience Newsletter*, from William Raub, associate director, NIH, and Joe R. Held, director of the division of research services, NIH, and a letter of response from Adrian Morrison and Peter Hand; "The Silver Spring Monkeys," *Tulane Medicine*, Autumn 1991; "Billy's Sad Life Fans Flames of Animal Rights Movement," by Jim Henderson, *Dallas Times Herald*, March 31, 1990. Also see "The Real Story of the Silver Spring Monkeys," a backgrounder prepared by Peter Gerone for public information. Gerone also discusses the transfer of the Silver Spring monkeys in a memo to John Walsh, chancellor of Tulane, January 5, 1987, Re: "Ownership of the Silver Spring Monkeys."

Background on Alex Pacheco and PETA comes from *The Washington Post* and *The New Yorker* articles. Both provide considerable background on Pacheco; the *Post* article discusses his work with the Sea Shepard and his confrontations over castrated cows at Ohio State University. The overall history of the animal rights movement is discussed in "Of Pain and Progress," in *Newsweek*, December 26, 1988.

The early history of the rise of animal activism is detailed in the Office of Laboratory Animal Care Report from University of California, Berkeley, September 1986, and in Susan Lederer's article, "Political Animals," *ISIS*, Vol. 83, 1992, pp. 61–79.

Background on Christine Stevens and the Animal Welfare Institute comes from personal interviews and two publications of the institute, *Beyond the Laboratory Door*, published in 1985, and *Animals and Their Legal Rights*, published in 1990. The examples of campaigns by In Defense of Animals are based on interviews with Roy Henrickson and coverage of IDA in "The Monkey Wars," *The Sacramento Bee*, November 1991. The section on Last Chance for Animals comes from a mailer sent out by the organization in the spring of 1993. Descriptions of the Animal Liberation Front are from "Animal Passion," *People* magazine, January 18, 1993, and a list of "Animal Rights Movement/ Illegal Incidents Summary" compiled by the National Association for Biomedical Research.

The Genarelli break-in is documented in PETA's video, "Unnecessary Fuss" and a transcription of the video, published in PETA News. Also *Science* magazine's account, "Lab Break-in Stirs Animal Welfare Debate," June 22, 1984, by Jeffrey L. Fox. A scientific explanation of the merits of Genarelli's work is contained in "Penn Work Was Vital," by Joseph Torg, *Philadelphia Inquirer*, June 8, 1984.

The story of the International Primate Protection League comes from personal interviews. Rowan's *Of Mice, Models and Men* (State University of New York Press, 1984) discusses the IPPL role in the ban on rhesus macaques from Bangladesh. The incident is also covered in IPPL newsletters, particularly the May 1982 issue, Vol. 9, No. 2.

Adrian Morrison spoke about his problems with the animal rights movement at the February 1993 meeting of the American Association of the Advancement of Science, in a speech titled "Scientific Freedom and Responsibility: A Retrospective." Morrison also spoke on "The Importance of Animals for Basic Biomedical Research" at an AAAS Symposium on the use of animals in research in February 1990.

The final chapter on the Silver Spring monkeys, the brain reorganization experiments, was published as "Massive Corticol Reorganization after Sensory Deafferentation in Adult Macaques," by Tim P. Pons, Preston E. Garraghty, Alexander K. Ommaya, Jon H. Kaas, Edward Taub, and Mortimer Miskin, *Science*, June 28, 1991. The results of those experiments are further discussed in "Famous Monkeys Provide Surprising Results," *Science* June 28, 1991; as well as "Brain Reorganization Puzzle in Silver Spring Monkeys," *The Journal of NIH Research*, September 1991.

Chapter 6.   The Peg-leg Pig

Interviews for this chapter include Peter Gerone and Robert Gormus, both at the Tulane Regional Primate Research Center; Frederick King, at Yerkes Regional Primate Research Center; Adrian Morrison, at NIH and the University of Pennsylvania; Frankie Trull, at the National Association for Biomedical Research; Suzanne Roy, at In Defense of Animals; Shirley McGreal, at IPPL; Martin Stephens, at the Humane Society of the United States.

A brief description of the impact of the Ebola crisis on primate importers can be found in "No Monkey Business," *The Journal of NIH Research*, Vol. 2, November 1990. The U.S. Centers for Disease Control inspection of Worldwide Primates is documented in a letter, sent March 22, 1990, to Matthew Block from Charles McCance, director of the Division of Quarantine at CDC. McGreal's letter to Gerone, concerning that inspection, is dated June 18, 1990. The lawsuit against McGreal was filed in the United States District Court, Southern District of Florida, under *Worldwide Primates Inc.* v. *Shirley McGreal*, Civil Action No. 90–2056–CIV–RYSKAMP. Court documents used as background for discussion include Gerone's response to McGreal's request for documents and McGreal's motion to dismiss the case, which cites Block's invocation of the Fifth Amendment in 49 out of 64 requests for information.

Background on the mangabeys was provided in discussions with Robert Gormus and Peter Gerone.

The coverage of the orangutan smuggling trial has been widespread. The case is *United States of America* v. *Matthew Block*, United States District Court, Southern District of Florida, Southern Division, No. 92–115–CR–KEHOE. McGreal reported on the case extensively in her newsletter, Vol. 19, No. 1, April 1992; Vol. 20, No. 1, April 1993. Also see "Animal Importer Admits Primate Conspiracy," *Science*, Vol. 259, February 26, 1991, p. 1256. (There is further discussion of the case in Chapter 11.)

Several journalists have reported on the full Ingrid Newkirk quote involving "a rat is a pig is a dog is a boy." They include Coleman McCarthy, *The Washington Post*, February 22, 1993; and Carolyn Fraser, "The Raid at Silver Spring," *The New Yorker*, April 19, 1993.

Peter Gerone's letter, complaining about the nitpicky inspection of his laboratory, was sent to James W. Glosser, administrator of the USDA's Animal and Plant Health Inspection Service, on April 21, 1989.

Suzanne Roy's memo on scientific dirty tricks is based largely on news reports. *Science* magazine reported on the role of biomedical researchers in linking animal rightists to the slaying in Tennessee in a brief article, "False Alarm from Vet Slaying?" in the March 9, 1990 issue.

The aggressive position of the Humane Society of the United States during the 1980s was noted in an article written by Frederick King, "Animals in Research: The Case for Experimentation," *Psychology Today*, September 1984, pp. 56–58.

Shirley McGreal's *IPPL Newsletter* covered the "AMA White Paper" on responding to animal activists extensively in the December 1989 issue. Further details are in her letter to the American Civil Liberties Union, sent September 4, 1989 to the organization's New York office. See also "AMA Asked to Seek Protection for Researchers from Political Interference by Animal Activists," *Journal of the American Medical Association*, Vol. 266, No. 4, July 24–31 1991.

Background on the National Association for Biomedical Research can be found in several documents published by the organization, including "1991 Highlights," the "1992 Annual Report," and a 1991 brochure celebrating "A Decade of Accomplishment." Christine Stevens, of the Animal Welfare Institute, discusses NABR's approach in a commentary in the journal *Nature*, September 1984, titled "Mistreatment of Laboratory Animals Endangers Biomedical Research." In an "Emergency" notice to its members, the Society for Animal Protective Legislation, also founded by Stevens, warned on August 24, 1990 that NABR was leading the effort to water down the regulations in the Animal Welfare Act. Frankie Trull's letter, applauding the final regulations, was reported in the *IPPL Newsletter*, Vol. 17, No. 3, November 1990. NABR's intentions to challenge any changes to those regulations are discussed in the NABR "Updates" of March 24, 1993 and May 19, 1993.

Biomedical research organizations are discussed in two articles in *The Scientist*: "Opponents Set 1993 Tactics for Animal Rights Showdown," Vol. 7, No.

2, January 25, 1993, and "World Laboratory Animal Liberation Week: Protests Fail to Weaken Scientists' Resolve," Vol. 7, No. 11, May 31, 1993. Both articles are by Ron Kaufman. The Americans for Medical Progress recruitment letter from Frederick King was sent on January 29, 1993. The Jewish task force letter from Americans for Medical Progress was an August–September 1992 mailing. The letter protesting Matthew Block's indictment was in *The Jewish Press*, February 26, 1993, 97, titled "Orthodox Jewish Father May Be Sentenced." The summer 1993 issue of *Earth Island Journal* lists Putting People First as an anti-environmental organization. Kathleen Marqueth's commentary on the White House Science advisor, John Gibbons, was published in *Fur Age Weekly*, March 1, 1993.

Discussions of education programs by groups on both sides of the issue are detailed in *The Scientist* articles outlined above.

Chapter 7. Hear No Evil

Interviews for this chapter included Susan Lederer, at the Pennsylvania State University; Larry Jacobsen, at Wisconsin Regional Primate Center, University of Wisconsin, Madison; Steven Kaufman, at the Medical Modernization Research Committee; Seymour Levine, at Stanford University; Jan Moor-Jankowski, at LEMSIP; Philip Byler, an attorney; Jim Mahoney, at LEMSIP; Frankie Trull, at NABR; Roger Fouts, at Central Washington University; Alex Pacheco, at PETA; and Shirley McGreal, at IPPL.

The story of Francis Rous Peyton and the *Journal of Experimental Medicine* can be found in Lederer's fascinating article, "Political Animals: The Shaping of Biomedical Research Literature in the 20th Century," *Isis*, Vol. 83, 1992, pp. 61–79.

Larry Jacobsen kindly provided printouts of some of the archives of "Primate Talk" so that I might have an understanding of the kind of conversations that are carried on the network. Other examples came from scientists around the country who use the network. The *International Directory of Primatology* was published in 1992, in part with funding from an NIH grant. It is divided into five sections, a listing of organizations that work with primates; a listing of field studies with primates; and sections on primate management groups, primate societies, and information resources.

The issue of secrecy in the new Animal Welfare Act regulations is covered in IPPL's *Newsletter*, in "Storm Over Housing Standards," November 1990, Vol. 17, No. 3. The opinion of Judge Charles Richey, U.S. District Court, District of Columbia, was issued February 25, 1993. The Animal Welfare Institute reported his ruling as "Judge Richey Rules in Favor of Decent Treatment for Lab Dogs and Primates," *AWI Quarterly*, Winter 1993.

The *Quantum Leap* story appeared in the Calendar section of the *Los Angeles Times*, "Quantum Leaps into Fray," by Howard Rosenberg, August 12, 1991. Trull's memorandum asking for researcher comment was sent out on

July 31, 1991; her own letter to Bellasario was written July 25, 1991. The head injury study, to which Moor-Jankowski refers in his letter, was "Head Injury in the Chimpanzee: Part 1. Biodynamics of Traumatic Unconsciousness," *Neurosurgery*, Vol. 39, August 1973, by Arub Ommaya, Paul Conrad, and Frank Letcher.

The Foundation for Biomedical Research reported that a "Biased Entry Appears in *Encyclopedia Britannica*," in its January–February 1992 *Newsletter*, Vol. IX, No. 1. Peter Gerone's letter of protest was sent to the chairman of the Board of Directors of the company, Robert Gwinn, on December 5, 1991. The letter from the American Society for Pharmacology and Experimental Therapeutics, applauding the changes in the dog entry, was set March 24, 1993 from the society's president, A. E. Takemori to the general editor of *Encyclopedia Britannica*, Robert McHenry. The author of the original dog entry, Michael Fox, discusses his changing attitude toward animal research in "A Change of Heart," reprinted in the National Anti-vivisection Society Bulletin from a 1987 piece in *New Age Journal*.

The section on suppression of viewpoints that challenge mainstream science is gathered both from interviews and other sources, among them "Censored: Faculty Who Oppose Vivisection," by Joan Dunayer, *Z Magazine*, April 1993, pp. 57–60. Stephen Kaufman provided a copy of his yet unpublished paper, "Animal Experimentation and the Big Lie," which details his experiences in trying to engage universities in debate over the animal research issue. Several sociologists have explored the ways in which scientists tend to avoid confronting the actual pain and death of animals in their care, among them Mary T. Phillips, "Savages, Drunks and Lab Animals: The Researcher's Perception of Pain," in *Animals and Society*, Vol. 1, No. 1, and Arnold Arluke, from Northeastern University in Boston, "Trapped in a Guilt Cage," *New Scientist*, April 4, 1992.

The story of the Immuno lawsuit against LEMSIP was covered in "Monkey Business: A Letter to the Editor Sparks a First Amendment Brawl," by John Strausbach, *New York Press*, March 6–12, 1991. Judith Reitman's book, *Stolen for Profit*, (Pharos Books, 1993) also discusses the case (pp. 179–80). Reitman notes, that in another book, *Make No Law*, New York Times columnist Anthony Lewis described the case as "the single most outrageous libel case—the worst abuse of the legal process." The case is also discussed in *Visions of Caliban*, by Jane Goodall and Dale Peterson (Houghton Mifflin, 1993). In his book, *Of Mice, Models and Men* (State University of New York, 1984), Andrew Rowan discusses the transfer of chimpanzees from LEMSIP to Southwest and the resulting ill feelings between Moor-Jankowski and NIH.

In preparing this chapter, I relied heavily on court documents. They included the formal opinion of the New York State Court of Appeals, January 30, 1989. Others were the brief filed on behalf of NABR, by attorneys Reed Smith Shaw & McClay, October 3, 1989; the affidavit filed in response by New York University, Floyd Abrams, attorney, October 13, 1989, which includes correspondence between Frankie Trull and David Scotch and the affidavit filed in response by Philip Byler, on behalf of Moor-Jankowski, on October 11, 1989. The amicus curiae brief, filed on behalf of the Sierra Club et al. by attorney Laura Mattera, includes transcripts from the deposition in which McGreal was asked about sexual favors and in which Moor-Jankowski complained about Nazi tactics and left the room. The lists of other amici curiae briefs filed in the case are also based on court records.

Chapter 8. The Salt in the Soup

This chapter is based largely on interviews with Jeff Roberts, at the California Regional Primate Research Center, University of California, Davis; Ron DeHaven, at the U.S. Department of Agriculture; Seymour Levine, at Stanford University; Deborah Fouts, at Central Washington University; Irwin Bernstein, at University of Georgia; Viktor Reinhardt, at Wisconsin Regional Primate Research Center, University of Wisconsin, Madison; Dee Carey, at the Southwest Foundation for Biomedical Research; and Christine Stevens, at Animal Welfare Institute.

The studies at the California Regional Primate Research Center, looking at housing of captive monkeys, were headed by Scott Line, a veterinarian who later left for Bowman Gray University in North Carolina. They include "Cage Size and Environmental Enrichment: Effects upon Behavioral and Physiological Responses to the Stress of Daily Events," by Scott W. Line, Hal Markowitz, Kathleen Morgan and Sharon Strong, in *Through the Looking Glass*, edited by M. A. Novak and A. Petto (American Psychological Association 1991). Also see "Influence of Cage Size on Heart Rate and Behavior in Rhesus Monkeys," by Scott W. Line, Kathleen Morgan, Hal Markowitz, and Sharon Strong, in the *American Journal of Veterinary Research*, Vol. 50, No. 9, September 1989; "Evaluation of Attempts to Enrich the Environment of Singly-Caged Non-Human Primates," by Scott Line, Hal Markowitz, Kathleen Morgan, and Sharon Strong, in *Animal Care and Use in Behavioral Research: Regulations, Issues and Applications*, edited by J. Driscoll (Animal Welfare Information Center, National Agricultural Library, 1989); and "Behavioral Responses of Female Long-Tailed Macaques to Pair Formation", by Scott W. Line, Kathleen N. Morgan, Hal Markowitz, Jeffrey A. Roberts, and Mike Riddell, in *Laboratory Primate Newsletter*, Vol. 29, No. 4, 1990, pp. 1–5.

Ron DeHaven provided much of the detailed explanation of how the USDA has attempted to respond to the requirements of the 1985 Animal Welfare Act. Details are also set forth in the act itself, the regulations published by the USDA in the *Federal Register*, and in the February 25, 1993 opinion of Judge Charles Richey. The International Primate Protection League published a detailed analysis of Richey's ruling in its April 1993 newsletter; Christine Steven's group, the Animal Welfare Institute, also reported on Richey's ruling in its winter 1993 newsletter.

The discussion of restraining devices comes from Ron DeHaven, Dee Carey, and Tom Gordon, director of the Yerkes field station, who worked on monkey studies in the 1960s which involved chairing monkeys and administering electric shocks.

Chapter 9.   Not a Nice Death

This chapter includes interviews with Roy Henrickson, at the University of California, Berkeley; Nick Lerche, at California Regional Primate Research Center; Henry McGill and Jonathan Allan, both at the Southwest Foundation for Biomedical Research, San Antonio, Texas; and Jan Moor-Jankowski, at LEMSIP.

The early history of the monkey AIDS model comes largely from the interviews. Other references include "The Long Shot," by Mark Caldwell, *Discover*, April 1993, in a special report titled "Why We Don't Have an AIDS vaccine." See also the May 28, 1993 issue of *Science*, "AIDS: The Unanswered Questions," and in particular, an article by Jon Cohen, "AIDS Research: The Mood Is Uncertain," in which top experts across the country discuss the issues that make the virus so difficult.

Background on retroviruses and how they work is also based, partly, on both "AIDS: The Unanswered Questions" and Cohen's "AIDS Research." In addition, see the book *The Science of Viruses*, by Ann Giudici Fettner (Quill Books, 1990).

In a series I wrote for *The Sacramento Bee*, also called "The Monkey Wars," November 1991, Andrew Hendrickx, director of the California Regional Primate Research Center, discussed the history of Accutane. Further references on Accutane include "FDA Advised to Limit Accutane Prescriptions," by Michael Abramowitz, *The Washington Post*, April 26, 1988; "Acne Drug as Dangerous as Thalidomide for Birth Defects," *Los Angeles Times*, October 14, 1988; and "Is FDA Study of Accutane Credible," by Michael Abramowitz, *The Washington Post*, April 25, 1988.

Betsy Todd's master's thesis on problems with animal models and AIDS is quoted by permission of the author. The full citation is *Animal Research and Aids*, Unpublished Master's Thesis, Columbia University School of Health, New York City, 1991.

In addition to Todd's study, reports on the use of chimpanzees in AIDS research, include "Drug Blocks HIV Infection in Chimpanzees," by Marilyn Chase, *The Wall Street Journal*, August 1, 1991, p. B–1, and "Two Chimps Immunized Against AIDS, Firm Says," by Robert Steinbrook, *Los Angeles Times*, June 2, 1990, p. A–29. Goodall and Peterson provide a critical analysis of chimpanzee AIDS research in *Visions of Caliban*, (Houghton Mifflin, 1993), which looks at the role of Dr. Robert Gallo in arguing for their use. Further discussion of the use of chimpanzees is included in scientific reports on the value of pigtail macaques.

The primary sources on pigtail macaques include "A Surprise Model Animal for AIDS," by Joseph Palca, *Science*, June 19, 1992; "Researchers Find Monkey That Can Contract AIDS," by Philip J. Hilts, *New York Times*, June 7, 1993; "Scientists Infect Monkeys with HIV," by Matt Crenson, *Dallas Morning News*, June 12, 1992; and "Too Much Monkey Business? Pigtailed Macaques and AIDS Research," by Carol Ezzell, *The Journal of NIH Research*, June 1993.

Genetic variation and evolution of AIDS viruses are discussed in the quarterly report of the Southwest Foundation for Biomedical Research, titled *Progress in Biomedical Research*. See, in particular, the winter 1992 edition, featuring "Out of Africa: Viral Evolution and AIDS," a report on the work of Jonathan Allan. Also see "The Future of AIDS," by Geoffrey Cowle, *Newsweek*, March 22, 1993, pp. 47–52.

The issues in developing a vaccine against AIDS are discussed thoroughly in the *Discover* piece by Mark Caldwell. Other references include "Researchers Discouraged about Prospects for AIDS Vaccine," by Daniel Q. Haney, Associated Press, June 17, 1988; "Inexplicable Results Throw a Monkey Wrench into AIDS Vaccine Research," by Carol Ezzell, *The Journal of NIH Research*, April 1992; and "Deadly But Elusive," by Mark Roth, *Pittsburgh Post-Gazette*, Monday, March 7, 1988, which explores the idea that the strains of AIDS viruses that grow in test tubes differ from those that grow, and kill, in living bodies.

Chapter 10. *Just Another Jerk Scientist*

This chapter is based on interviews, some quoted, some used as background, with Jonathan Allan, at the Southwest Foundation for Biomedical Research; Nick Lerche and Jeff Roberts, at the California Regional Primate Research Center; Tom Kziasek, of the special pathogens division of the U.S. Centers for Disease Control; and Seymour Kalter and Richard Heberling, at the Viral Reference Laboratory, San Antonio, Texas.

The point of view of transplant surgeons is taken mostly from press accounts. I interviewed Leonard Bailey myself, in June 1991, for *The Sacramento Bee* series, "The Monkey Wars," published in November of that year. Starzl's work has been so public that his opinions are widely known and widely quoted. Among the references used in covering his work was the University of Pittsburgh Medical Center's press kit on xenotransplantation, which includes statements on baboon viruses, on anti-rejection drugs, a biography of Starzl, case studies of both baboon transplant cases at Pittsburgh and a history of xenotransplantation. In addition, press reports of the transplant surgeries were studied. They included "Terminally Ill Man with Damaged Liver Receives Transplant from Baboon," by Lawrence K. Altman, *The New York Times*, June 29, 1992; "Transplant Protest Draws Heat from Liver-Disease Victim," Associated

Press, *San Antonio Light,* July 1, 1992, which reports on Robert Winters' exchange with animal advocates; "Life Debate: Research Scientists Defend Use of Baboon Liver in Transplant," by Bob Benis, *Austin American–Statesman,* July 6, 1992, which discusses the fact that Pittsburgh did not notify Southwest about its need for a transplant baboon in advance. Starzl's quote about his commitment to patients comes from "Animal Rights, Human Lives," by B. D. Cohen, *Newsday,* July 9, 1992, and the ethics of using baboons as human parts are further discussed in another Cohen piece for *Newsday,* "A Baboon Dies, A Man Lives," July 7, 1992.

Herpes B is a heavily studied virus. I am indebted to Julia Hilliard, of the Southwest Foundation, and Seymour Kalter and Richard Heberling, of the Virus Reference Laboratory in San Antonio, Texas, for providing me with much background information on the virus. References include "B Virus Infection of Primates in Perspective," by Kalter and Heberling, *Lab Animal,* April 1989; "Biology of B Virus in Macaque and Human Hosts: A Review," by Benjamin Weigler, *Clinical Infectious Diseases* (The University of Chicago, 1992), which discusses the fears about B virus infection in polio vaccines; "B Virus; Herpesvirus Simiae: Historical Perspective," by Amos E. Palmer, *Journal of Medical Primatology,* Vol. 16, 1987, which reviews all cases of human B virus infection until that point; "B Virus Infection in Humans; Epidemiologic Investigation of a Cluster," by Gary Holmes, Julia Hilliard, et al., *Annals of Internal Medicine,* June 1990, which reviews the serious outbreak of B virus in Florida in 1987. The first article on B virus is "Acute Ascending Myelitis Following a Monkey Bite, with Isolation of a Virus Capable of Reproducing the Disease," by Albert Sabin and Arthur Wright, *New England Journal of Medicine,* Vol. 59, 1934.

Descriptions of SV40, the papovaviruses, and the many infections carried by wild primates can be found in Manfred Brack's book, *Agents Transmissible from Simians to Man* (Springer-Verlag, 1987). Also see "Primates in Viral Oncology," by Richard L. Heberling, in *Viral and Immunological Diseases in Nonhuman Primates* (Alan R. Liss, 1983); "Primate Viral Diseases in Perspective," by S. S. Kalter and R. L. Heberling, *Journal of Medical Primatology,* Vol. 19, 1990;. and "An Overview of Biohazards Associated with Nonhuman Primates," by Elizabeth Muchmore, *Journal of Medical Primatology,* Vol. 16, 1987.

The best description of the Ebola outbreak in the United States can be found in Richard Preston's "Crisis in the Hot Zone," *The New Yorker,* October 26, 1992. Other references include "Scientists Trace Ebola Virus's Deadly Path," by D'Vera Cohn, *The Washington Post,* December 11, 1990; "Outbreak of Fatal Illness Among Captive Macaques in the Philippines Caused by an Ebola-Related Filovirus," by Curtis G. Hayes et. al., *American Journal of Tropical Medicine,* Vol. 46, 1992; "Import Rules Threaten Research on Primates," *Science,* June 1, 1990; and "Not Enough Monkey Business," *Science,* October 26, 1990. I refer also to CDC's "Dear Primate Importer" letter of March 15, 1990, warning of health risks, and the agency's "Dear Interested Parties" letter of August 27, 1991, detailing the results of the quarantine to protect against Ebola importation.

In *Rolling Stone*, March 19, 1992, writer Tom Curtis raises the issue of whether the polio vaccines, given in Africa, accidentally began the AIDS epidemic. Curtis also provides background on SV40 and B virus. The article is titled "The Origin of AIDS." The ideas are further explored in "The Mysterious Origin of AIDS," by Jared Diamond, *Natural History*, September 1992, which discusses the history of the early experiments which injected people with monkey blood as well. *The Journal of NIH Research* also explored the possibilities in "Fact or Fiction: HIV and Polio Vaccines," by Nancy Touchette, September 1992.

The Journal of NIH Research also published one of the better analyses of the question of transplanting viruses, along with baboon livers, in "Hope or Horror? Primate-to-Human Organ Transplants," by Rachel Nowak, in September 1992. The listing of baboon viruses screened for by Pittsburgh researchers comes from unpublished data. Two articles on foamy viruses used as background include "Isolation of a New Foamy Virus from an Orangutan with Encephalopathy," by Myra McClure et al., presented at the 10th Annual Symposium on Nonhuman Primate Models for AIDS, November 17–20, 1992, San Juan, Puerto Rico; and "Human Spumaretrovirus-Related Sequences in the DNA of Leukocytes from Patients with Graves Disease," by Sylvie Lagaye et al., in the *Proceedings of the National Academy of Sciences*, November 1992.

The experiment mixing baboon and mouse-sarcoma viruses was formally reported by Kalter and Heberling as "Primate Endogenous Viruses: Their Role in Oncogenesis and as Biohazards" at the Joint WHO/LABS Symposium on Standardization of Cell Substrates for the Production of Virus Vaccines, December 1976. It had received earlier coverage in *The New York Times* by H. M. Schmeck on May 28, 1976. The letter of protest in *Science* magazine appeared in the July 23, 1976 edition. The letter was signed by Lawrence A. Loeb and Kenneth D. Tartof, of the Institute for Cancer Research in Philadelphia. And Kalter's research was reported on in the winter 1991–92 edition of *Covert Action*, in an article by Richard Hatch titled "Cancer Warfare."

Chapter 11. The Last Mangabeys

This chapter is based largely on a series of interviews with Dr. Robert Gormus, of the Tulane Regional Primate Research Center. Others interviewed include Peter Gerone and Frank Cogswell, also at Tulane; Milton April, at NIH; Tom Gordon, at Yerkes Regional Primate Research Center; Jonathan Allan, at Southwest Foundation for Biomedical Research; Nick Lerche, at the California Regional Primate Research Center; and Shirley McGreal, of IPPL.

The information on the breeding colonies at the NIH-supported primate centers comes from discussions with administrators of the program, including April, Leo Whitehair, and Michael Dukelow. The background on how Yerkes is handling its great apes comes from Tom Gordon, director of the Yerkes Field Station.

The prices on monkeys are taken from a Charles River price list.

Background on tropical deforestation comes from a number of sources, among them *The Deluge and Ark,* by Dale Peterson (Avon Books, 1991), which explores the issues of endangered primates. A further reference was *Primates: The Road to Self-Sustaining Populations,* edited by Kurt Benirschke (Springer-Verlag, 1986), in particular Chapter 8, "The Conservation Status of Nonhuman Primates in Indonesia," by Kathleen MacKinnon, which provides very specific numbers on individual primate habitats for orangutans, gibbons, and crab-eating macaques. Other background came from "Action Plan for Conservation of Indonesian Primates, 1992–1996," by L. H. Prasetyo, I. Sugardjito, and R. R. Tenaza, *Primate Conservation,* (the Newsletter and Journal of the IUCN/SSC Primate Specialist Group), No. 9, December 1988, which was sponsored by Fort Wayne Children's Zoo, Conservation International, and the IUCN/SSC Primate Specialist Group's "Action Plan for Asian Primate Conservation: 1987–1991."

The monkey import totals were taken from documents provided by the British Union to Abolish Vivisection.

The estimates of mortality among trapped monkeys are widely known in scientific circles and are based on discussions with reputable dealers, particularly Chuck Darsono, who has worked with the University of Washington in setting up captive breeding colonies for crab-eating macaques in Indonesia.

The death rate among African Green monkeys, imported into the United States in 1978 and 1979, is cited in the January 1981 *Newsletter of the International Primate Protection League,* under "IPPL Uncovers High Shipment Mortality of Vervet Monkeys." The deaths of 110 crab-eating macaques imported by Matthew Block was also noted in the IPPL *Newsletter,* December 1992, under the title "Dead on Arrival—110 Times Over." The *Miami Herald* Sunday magazine, "Tropic," also covered the story; see "The Monkeys' Curse," by John Dorschner, November 29, 1992.

The story of the impact of polio research on rhesus macaques of India is covered in Andrew Rowan's book, *Of Mice, Models and Men* in the chapter on primate research. Charles Southwick, of the University of Colorado, is one of the foremost experts on Indian macaques, and other background came from his "Utilization, Availability and Conservation of Nonhuman Primates for Biomedical Programs," presented at the International Symposium on the Use of Nonhuman Primates in Biomedical Research, New Delhi, India, 1977, published by the Indian National Science Academy, as well as a brief article in *National Geographic,* titled "The Outlook Brightens for India's Monkeys," November 1969, which cites Southwick's surveys of rhesus macaques.

The debate over using pigtail macaques is covered in the *Journal of NIH Research* article "Too Much Monkey Business?" by Carol Ezzell, in the June 1993 issue. Bowden's letter of response was sent to the magazine's Advice and

Dissent section on June 24, 1993. Bowden and Orville A. Smith, also of the Washington Primate Center, discuss the issue in general in "Conservationally Sound Assurance of Primate Supply and Diversity," *ILAR News*, Vol. 34, No. 4, Fall 1992.

The story of the Bangkok Six orangutans was described in detail, both on the BBC video *Ape Trade* and in "Orang Odyssey," by Jessica Speart, in *Wildlife Conservation*, November–December 1992, a publication of the New York Zoological Society. The Miami weekly publication, *New Times*, also described the tensions between McGreal and Block in an account that preceded the trial, called "The Primate Debate," by William Labbee, November 20–26, 1991. Block was indicted on smuggling charges on February 20, 1992. His trial received wide press attention and detailed coverage in IPPL newsletters, notably "Indictment in 'Bangkok Six' Orangutan Case," Vol. 19, No. 1, April 1992; "Criminal Case against Block," August 1992; and "Block Gets Jail," Vol. 20, No. 2, August 1992. It is in the latter, which quotes lengthy sections from the court proceedings, that both Block's request to carry a gun and the comments of Dr. Chambers appear. The comments from Robert Sapolsky are taken from a copy of his letter, sent to Judge James Kehoe, on March 29, 1993, from Stanford University.

The story of Robert Gormus's mangabey hunt is based primarily on personal interviews. The pseudonym John Bryant is the author's creation. Dr. Gormus did provide documentation as well, including some sections of his diary and the Federal Fish and Wildlife Permit for the mangabeys, issued December 12, 1988, No. PRT–712638. Personal correspondence from Gormus, trying to salvage the program, includes letters in the spring of 1993 to the Armed Forces Institute of Pathology and to NIH. The NIH response, announcing that the grant would not be funded, was mailed on June 29, 1993, from Darrel D. Gwinn, of the Division of Microbiology and Infectious Disease at NIAID. The letters provided many of the details of the actual experiments in progress. Basic background on leprosy also came from the 1992 edition of *Encyclopedia Americana*. Gormus learned in April 1994 that NIH was likely to provide enough money to finish the leprosy vaccine tests. At that point, he had been without funding for 18 months.

Chapter 12.   One Nation

This chapter is based on interviews and synthesis of information reported throughout the book. Interviews for this chapter included Tom Gordon and Frans de Waal, both at the Yerkes Regional Primate Research Center; Irwin Bernstein, at the University of Georgia; and Duane Rumbaugh, at Georgia State University.

INDEX